Adam Nicolson is the auth... bestselling books on hist... *Sissinghurst*, *Sea Room* and *When God Spoke English*. *The Mighty Dead* was longlisted for the Samuel Johnson Prize, and he has won the Somerset Maugham Prize, the WH Heinemann Prize and the Ondaatje Prize. He has written and presented many series for television, most recently *The Century That Wrote Itself*, about life and writing in the 17th century. He is a Fellow of the Royal Society of Literature and lives in Sussex with his wife and family.

Smell of
Summer Grass

Pursuing Happiness
at Perch Hill

ADAM NICOLSON

WILLIAM
COLLINS

William Collins
An imprint of HarperCollins*Publishers*
1 London Bridge Street
London SE1 9GF
WilliamCollinsBooks.com

First published by Harper*Press* in 2011

This William Collins paperback edition published 2015

1

Copyright © Adam Nicolson 2011

Adam Nicolson asserts the moral right to be identified as the author of this work

Parts of this book were previously published in PERCH HILL (Robinson Publishing, 1999)

A catalogue record for this book is available from the British Library

ISBN 978-0-00-810472-6

Typeset in 12/15pt Minion by Palimpsest Book Production Ltd, Falkirk, Stirlingshire
Printed and bound in Great Britain

In memory of Simon Bishop
1958–2009

Acknowledgements

The following images are reproduced with many thanks:

Section I page 2 top	Jeremy Newick
Section I page 2 bottom	Andrew Palmer
Section I page 5 bottom	Jonathan Buckley
Section I page 8 top	Alexandre Bailhache
Section II page 2 top	Jonathan Buckley
Section II page 3 bottom	Alun Price
Section II pages 4-5	Jonathan Buckley
Endpapers	Ricca Kawai.

Large parts of this book first appeared in the *Sunday Telegraph Magazine* between 1995 and 2000 and two-thirds of it between hard covers in *Perch Hill: a new life*, published by Constable Robinson in 1999. I would very much like to thank Charles Moore, Alexander Chancellor, Aurea Carpenter and Nick Robinson, my various editors in those places, for all their help and guidance. This book takes the Perch Hill story on another full decade and looks again, with a slightly longer perspective, at those early days on the farm. This time I would again like to thank my editor Susan Watt, who has stood by me through thick and thin over many years, and my dearly valued agent Georgina Capel.

Nothing at Perch Hill could ever have happened without the people who work there and I would like to acknowledge with enormous and deeply felt thanks the difference which Tessa Bishop, Colin Pilbeam, Bea Burke, Angie Wilkins and Ben Cole have all made to our lives. Nothing, in my experience, can match the feeling which a joint and shared attachment to a place can give.

Almost needless to say – as anyone who reads these pages will discover it soon enough for themselves – the part of Sarah Raven in this story is not far short of the role played by gravity in the universe.

Adam Nicolson
Perch Hill 2014

Contents

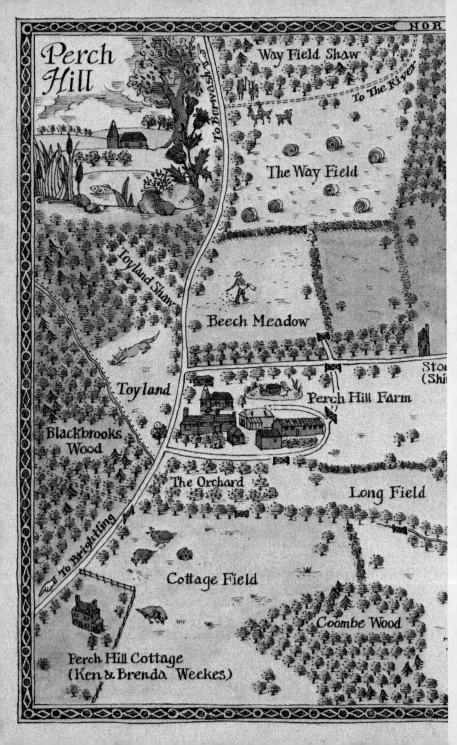

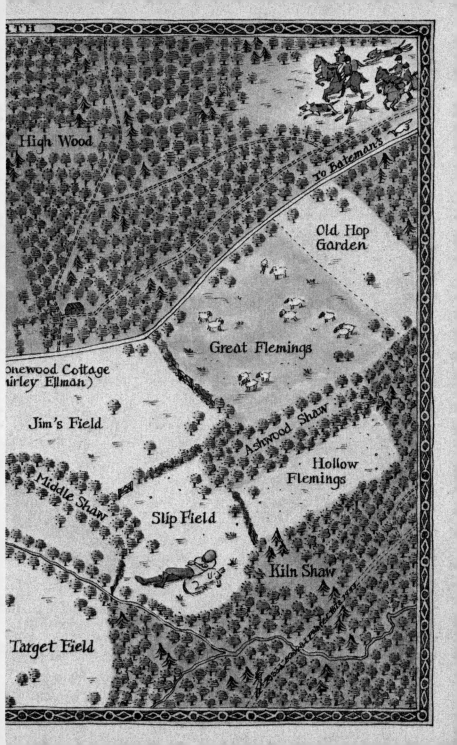

Part One

BREAKING

The Bright Field

IF I think of the time when we decided to come here in 1992, it is a backward glance into the dark.

A summer night. I am walking home from Mayfair, from dinner with a man I fear and distrust. He is my stepfather and I burp his food into the night air. It is sole and gooseberry mousse. His dining-room is lined in Chinese silk on which parakeets and birds of paradise were painted in Macao some years ago. The birds have kept their colours, they are the colour of flames, but the branches on which they once sat have faded back into the grey silk of the sky. On the table are silver swans, whose wings open to reveal the salt. The Madeiran linen, the polished mahogany, the dumb waiter: it's alien country.

My stepfather and I do not communicate. 'It's only worth reading one book a year,' he says. 'The trouble with this country is the over-education of the young.' 'Calling a parcels service "Red Star" is a sign of the depth of communist influence, even now, in England.'

Nothing is given. I leave the house to walk across London to somewhere on the edges of Hammersmith, where I am living with Sarah Raven, the woman for whom, a few months previously, I have left my wife. That is a phrase which leaves me raw. Sarah has gone somewhere else this evening, to have dinner with friends, and won't be back until midnight or later. I have left as early as I can from the Mayfair house and think 'Why not?'

A warm night. A walk through London and its glitter in the dark, to expunge that padded house and all its upholstered hostilities.

There are whores in the street outside, bending down like mannequins to the windows of the slowing cars. Some of the women are so tall and so sweetly spoken they can only be men, with long, stockinged calves and a slow flitter to their eyes. One fixes me straight. 'Looking for company?' she asks. 'Thank you, not tonight, thanks,' and we move on.

I walk down Park Lane, where the cars are thick and the night heavy. The lights from the cars blip, blip, blip through the dark. Life is hurried. I pass the Dorchester and the Hilton, down to the corner where the subway drops into ungraffitied neon. Down and up and down again to the upper parts of Knightsbridge. Along there the windows gleam. A friend of mine has opened a shop in Beauchamp Place. It is lacquered scarlet inside and beautiful, with black Japanese furniture standing on the hardwood floor.

On and out, increasingly out, away from the polish of the glimmer-zone, made shinier at night, to the open-late businesses of South Ken, where Frascati and champagne stand cooled in ranks behind glass and the Indian at the till leaves a cigarette always smoking on the ledge beside him. On, out, westwards, where occasional restaurants are all that interrupt the domestic streets now tailing into dark. The pubs have shut. It is nearing midnight.

For a stretch the street lights are broken, perhaps two in a row, and it is darker here. I think nothing of it. My mind is on other things. On what? I was thinking of a place where I have been happy, some kind of mind-cinema of it flicking through my brain: sitting back on the oars in the sunshine in a small boat off the coast of some islands in the Hebrides, where black cliffs drop into a sea the colour of green ink and the sea caves at their feet drive 50 yards or more into pink, coralline depths.

That night I was thinking of these things, of hauling crabs from the sea, scrambling among the hissing shags and peering down the dark slum tunnels where the puffins live, lying down in the long grass while the ravens honked and flicked above me and the buzzards cruised. My mind was away there that night.

I must fight a reluctance to describe what followed. I am wearing a suit, an Italian suit I have had for years, with turn-ups to the trousers and pointed tips to the lapels. It is a sharkish, double-breasted thing. The Mayfair whores had seen a businessman inside it and so, I suppose, did the three youths, late teenagers, in the Lillie Road.

The heels of my shoes were striking the pavement too hard, like flints. I tried to soften them by treading on the balls of my feet. Two of the boys were on the inside of the pavement next to the wall. I did not look at them. The other was on the kerb. I walked between the three as though through an alley and adrenalin shocked into me as I saw their eyes go white in the unlit street. I saw the kerbside boy nod at the others. I thought how contemptible was my *Daily Mail* fear of these people. I was already beyond them, and relieved, when my eyes and mouth stung and burned and there were hurried hands under my armpits pulling and pushing me into the mouth of a passageway leading off the road. My body had hunched over as the ammonia came into my face – bleach squirted from a lemon squeezer – and they knelt me on the gritty pavement, as though I were being unpacked, a bale of stuff, my body and suit a pocketed rucksack, all hurry and hard fingers against my ribs. I said nothing. I tried to get up but they rubbed the bleach into my eyes, oddly without violence, in the way you would pull back on the chain of a dog, simply a control.

I was not a person but a suit with pockets. I was being fleeced, in the way a shepherd might fleece a sheep. My assets were being stripped. I knelt with the grit of the pavement pricking in my cheek while they looked for money and objects in the suit that was no longer mine. They were robbing the suit. The bleach had emptied it of a person, I could not help but regard from a distance this odd, disembodied theft. I was in pain but the burning in my eyes and mouth seemed unrelated to this professional going-over of my clothes, not my clothes, *the* clothes, *some* clothes.

They left, up the passageway. I lay for a moment on the concrete slabs, excited by the reality of what had happened. My eyes were blurry and my tongue was ulcered and raw. I can taste and smell the ammonia now, years afterwards, a chemical thickness to it, a fog of fumes rising from my mouth into my nose. I got up. I dusted the suit off; it was torn. I walked down to the North End Road. There was a fish-and-chip shop open there. I went in and asked the man behind the counter if I could wash my face in his basin. He looked at me. His apron was up around his armpits. 'We've been messing about a bit, have we?'

'No,' I said. 'I've just been fucking attacked.'

He showed me a room which had a basin and a towel in it. I washed there, deep in the water, holding the water to my face and eyes, wanting to wash the pain away, and the taste of the bleach, and the furry, clogged thickness on my tongue, but feeling, more than anything, broken, hopeless, at the end of a long and hopeless trajectory which, for many months and even years, had curved only down.

I walked to the house. It wasn't far. I sat down on the doorstep. I said to myself I was fine. But I knew I wasn't and eventually ended up in hospital where, at three in the morning, a doctor hosed the ammonia from my eyes, holding them open with his rubber-gloved fingers one by one, so that the water would sluice around the recesses of the eye. By pure chance, the doctor told me, precisely the same thing had happened to him the year before. Some Spurs fans had set upon him, squirted his eyes with bleach, robbed him and left him feeling blurry like this on the pavement. It was his way of consoling me, I suppose.

Only later, in Sarah's bed, deep in the night, with the grey-yellow wash of the London street lights leaking around the edges of the curtains, did I allow myself to cry, to sob out all the held-back reservoir of humiliation and failure whose dam the mugging had broken.

It was not the attack itself for which I wept and sweated that night but everything of which it seemed, however irrationally, a culmination: the failure of my first marriage the year before, my guilt at my own part in that failure, the effect my leaving would have on my three sons by that marriage, the failure or near-failure of a business I had been involved with for five years, which I had also abandoned, unable to work properly any longer, leaving it in the hands of my cousin and co-director at the one moment he most needed my help. On top of that, a book I had been trying and failing to write had finally collapsed in exhaustion and uncertainty. If I had been a horse I would have been shot. I should have been shot. I had broken down.

The mugging was a catalyst not of change, but of paralysis. I scarcely moved for three months. I lay in bed. Sarah went alone to work and to parties. I saw in her face a terror of what she had allowed into her life. I let everything about me – my own work, my sense of self-esteem, any idea of care or responsibility for others – fall away. Nothing meant anything to me. I could make no decisions. When I met people I knew, they looked into my face as though something were missing there. I woke up tired. I spoke more slowly than before. I saw a psychotherapist and told him that I felt like a sooted chimney, nothing but a dusty black hollow cylinder inside my skin. I felt that my breath polluted the air around me. I dreamed of my children. One night we were walking in a rocky place like Crete. 'I am sorry,' I told them. 'I must leave you behind,' and without waiting for an answer set off up the side of a mountain which reminded me of Mount Ida, its dry, limestone bulk, its sterility, its demand to be climbed. I arrived at the chapel on the summit, a place of bare rock, and slumped down beside the walls, my face in my hands, my body with every muscle slackened, every limb like a bone in a bag. When I looked up, I saw the three boys coming towards me, easily moving up on to the final rise, a bobbing movement, alive, lightened, untaxed by the journey on which I had deserted them. 'Why do divorced men become obsessed by their children?' I heard a woman ask. I could have told her: because they watch them from what seems like the far side of death.

In the face of all that, Sarah was life itself. I had met her on holiday together with a few friends. She knew my sister Rebecca and I still remember every minute of those first mornings with her. She was strong and fearless. She took control. She arranged things. I told jokes to make her laugh and she laughed with her whole head thrown back and her throat open. She didn't take any nonsense. She raced me downhill – we were skiing – and smoked on the lifts back up. She loved the west coast of Scotland and a half-abandoned house in chestnut woods in the valley of the Tarn. She was a doctor. She always voted Labour. She wore glamorous printed silk shirts from a company called English Eccentrics. She played with her long red-brown hair while talking to me. She was the natural focus of everyone around her. There was no side or twist to her: she was what she seemed to be. She could drink for England. She seemed to like me. She loved wild flowers. She never read a book. As she pointed out to me, she had beautiful long legs, very good for walking. She was in love with the cooking of the Veneto, which she had learned as a girl. Above all she had an appetite for living. She did not seem defeated. She looked not exactly like the future but like someone with whom and alongside whom the future was full of glow and richness. Life was full for her, not as an abstract idea – nothing intellectualized here – but as a reality which involved things, food, work, happiness, children, nature, gardens, beauty. She was the substance of life.

And so we fell in love – weeks and months of looking forward to seeing her and being with her, of being enlivened by her teasing, warm, loving presence. And with that, folded in with it, my own grief and despair at what had happened. I have never known things at the same time be so beautiful and so dark.

From her house in London, Sarah and I began to search for a refuge, however naïvely and hopelessly that idea was conceived. It stemmed from no more than a belief in the pastoral. 'Are not these woods more free from peril than the envious court?' a figure in *As You Like It* asks the surrounding company. I knew in the past I had been happy in rural places. I knew, or thought I knew, that a rural place would soothe this crisis. I knew, as I walked out in the streets of London, that there was no solace there. Every surface was dead in my eyes. My mind returned constantly to those islands in Scotland which I had been thinking of on the night of the attack. For 15 years I had owned them. My father had bought them 50 years before for £1,200 and he gave them to me when I was 21, as I was to give them to my son Tom when he was 18. Cynics have said that all this was for tax reasons, but it wasn't. I think my father gave them to me because, as a very young man, he had felt enlarged and excited by the ownership of a place like that, by the experience of being there alone or with friends, away from the thing that Auden called 'the great bat-shadow of home', the enclosing, claustro-phobic, involuntary oppression of a parental place, which makes a bawling, complaining infant of you. He wanted, I think, to give that same enlargement to me, as I do to my own children.

It worked and the gift was this: memories of weeks there, storm-battered, sun-stilled, on which I continue to draw every day of my life. I know those islands yard by yard, I know the places to clamber up and slither down, I know the particular corners where the pair of black guillemots always nests or where the bull seal hauls himself out on the seaweedy rock, I know where the fish congregate in the tidal streams or where the eddies riffle off a nose of lichened basalt and throw your dinghy out in a sudden curving arc towards the Lewis shore. I know the natural arch where the seals swim and where the kelp gathers in an almost Ecuadorian sun-barred forest beneath your coasting hull.

I was essentially shaped by those island times. Almost everything else feels less dense and less intense than those moments of exposure. The social world, the political world, the world of getting on with work and a career, all those were for ever cast in a shadow by the raging scale and seriousness of my moments of island life. That intimacy with the natural makes the human seem vacuous.

This may be straight Wordsworthianism and I would want to disown it in favour of a less monolithically obvious thing, a glitteringly complex attitude to nature which shimmered like an opal compared with my all-too-single basalt slab. But I can't. I know nothing bigger or finer than the feeling that all barriers are down and a full-blown flood is running to and fro between you and the rest of the world.

I know all these things and treat them as my touchstones and my yardstick. Is this life, I always ask, as good as that? Does this place measure up to that? That is the fixture; everything else can only eddy around it.

We began to search for somewhere that might be the equivalent of all that, a place which in its own terms could be an island, around which the cord could be drawn, and where life could in some ways be hidden, or even innocent. It was the search for an Arcadian simplicity in which crisis and breakdown did not and could not occur. Fantasia you might say, but it had then an urgency and reality stronger and more concrete than anything else in the world around us. There was no sense, it seemed to me, of 'getting away'. There was no desire to enter a capsule or satellite suspended above the earth. It felt, if anything, the very opposite of that, a burrowing in, a search for a bed in which the covers could be drawn up and over us. It was, I now see, these many years later, a search for a womb, a place in which you could be protected from damage. It was an infantile need and ferociously demanding because of that.

We roamed England with the template in our minds. It seems curious now that this search might have landed almost anywhere, that anywhere might have provided the bucket into which the love could have been poured. Dorset, Devon, Somerset, Shropshire, Herefordshire, Oxfordshire, Norfolk, Suffolk, east Kent: all for a time became the zone in which safety might be found. It looks pathetic now, the two of us, in the white 2CV we had at the time, poking about like moles for a burrow, living with a private intensity the common stuff of rural estate agents' offices.

I had no perspective on what we were doing, or at least suspended any perspective I might have had. We were the first to do this. Of course we weren't – we were the last, the heirs and successors of a line that goes back at least to the Roman love affair with the suburban villa, perhaps beyond that to the first urban civilizations of the Near East, where the concentrated demands and sophistication of city life produced, even at the beginning, a dream image of the garden place, the paradise, in which the realities did not impinge, where the commercial and competitive structures of the city were absent. Is Genesis itself, I now wonder, a symptom of a disenchanted urbanity?

I had no desire to delineate, let alone puncture, the bubble. I needed its insulation and a belief in its power and reality. For years I had kept in my mind, as a sort of mantra, a poem by R.S. Thomas:

> I have seen the sun break through
> to illuminate a small field
> for a while, and gone my way

and forgotten it. But that was the pearl
of great price, the one field that had
the treasure in it. I realize now
that I must give all that I have
to possess it. Life is not hurrying

on to a receding future, nor hankering after
an imagined past. It is the turning
aside like Moses to the miracle
of the lit bush, to a brightness
that seemed as transitory as your youth
once, but is the eternity that awaits you.

('The Bright Field', 1975)

Thomas, a parish priest in the Lleyn peninsula in Gwynedd, in
the north-west corner of Wales, is playing a fugue on the words
of both Exodus and the Gospels. The Authorised Version does
indeed speak of Moses 'turning aside' to the burning bush; Christ,
talking to the Apostles, describes the Kingdom of Heaven both
as the field with the treasure in it and as the pearl of great price.
Both, curiously perhaps to us who dissociate so firmly the religious from the financial, use the language of money and
merchants. The conditions of paradise, in Christ's own words,
can be bought. If only estate agents had cottoned on to this!
Sell all that you have, money for paradise, the pearl of great
price, life is neither hurrying nor hankering: turn aside to the
eternity that awaits you. Heaven is waiting, the paradise womb;
only look for it and you will see the bush alight beside you. The
brightness of youth can be once again to hand.

I look at myself then, nearly twenty years ago, and scarcely
recognize the man I was: driven by a hunger for authenticity, for
a place in the world that did not seem compromised but was
somehow, in reality, almost heavenly. And I feel both nostalgia
for that man – for the simplicity of his proposition – and utterly

removed from him, as if he were some remote cousin, sharing a few common traits, and with something of a shared history, but with an attitude to the world that seems to know about almost nothing but himself. His need for happiness was so powerful that it erased nearly everything around him. It was as if he were walking through the world surrounded by a halo of need.

My own financial state was catastrophic. I was scarcely in a condition to work, or to look for work. A kind of vertigo gripped me whenever I tried to write anything. I was producing occasional pieces for the *Sunday Times* but they were ground out like dust from a mortar. All fluency had gone. I wrote a book of captions to photographs of beauty spots. I ghosted another on how to take landscape photographs. I researched the illustrations for a book on evolution. Apart from the islands in Scotland, which I would sell only if death were the other choice, I owned nothing. I had given my house and its contents to my first wife. I was paying her virtually everything I earned. At times I didn't earn enough in the month to pay her what I had promised and Sarah, from her earnings as a doctor, made up the difference. Sarah, who had inherited a little money, at least owned her house in London, all save a small mortgage, and that was the lifeboat. That house in west London, with its three bedrooms and a sliver of a garden, would take us where we needed to go.

In that search for a place, nothing much was working. House after house was wrong, wrong in feeling, wrong in its situation, wrong in its price. The more the vision glowed, the less the places we saw came up to it. There were houses drowning in carpets and improvement, their souls erased. There were others in which the road was too near. Others too dour, many too far from Cambridge where my sons were living with their mother. Both Sarah and I felt an instinctive aversion to arable parts of England. We needed grass and wood, the Arcadian savannah, the lit bush, the field illuminated for a while. I knew that when we found the place it would say 'I am here and I am yours. I am the place.' It was a dark time.

Only at the edges, like jottings and little coloured drawings in the margins of a text, were there any points of light: the finding of wild daffodils one day in a Dorset wood; swimming one warm and languorous evening off Chesil Beach where the swell rolled in like a lion's stretch and yawn, over and over, a long slow growling from the shingle; a weekend on the Lizard in a sea of thrift; dawn on the Helford River, anchored in a boat between the woods, where the night rain dropped off the outstretched leaves into water as still as oil. These were bright fields too, illuminations in their way.

Why should it be that beauty can for an instant make sense of a world in which nothing else does? I am not sure. It is something I believe in without knowing why. Maybe it is simply a recognition of pattern, a concordance between you and the world. Here in a chance beautiful thing is something given, neither engineered nor sought, neither curiously made nor elaborately framed, but dropping as a bead of meaning out of a meaningless sky. Its value, its weight, is in your own recognition of its beauty. There is something naturally there which you naturally recognize as good. The ability to see that beauty is a sign that the world is not an anarchy of violence and destruction. You belong to it and it belongs to you.

That was as near as I could come to understanding why I wanted to live and be in a place that seemed beautiful. The world around you in such a place would constantly touch you and speak to you. It would become an existence thick with understanding and that sense of crowding intelligibility might be almost social in its effect, as though you were actually joining the community of the natural. This, in my loneliness and guilt, became a kind of consolation too and I held on to it as a kind of flag of hope, a thought with which I could identify and salvage what remained of my self. There's a sentence in one of Coleridge's notebooks for 1807, when, also with a broken marriage and a career in ruins, he was staying on a farm in Somerset. A ragged peacock walked the yard: 'The molting Peacock with only two of his long tail feathers remaining, & those sadly in tatters, yet proudly as ever spreading out his ruined fan in the Sun & Breeze.' That was me with my faith in redemption by beauty, like the battle-shot colours of a regiment held up to the last.

I was at work in London – I say at work; I was sitting at my desk, looking at the screen, drinking a cup of coffee, considering from a dead mind the identity of the next possible word – and the phone rang. Sarah, in a coinbox, in a pub, breathless: 'You must come. This is the place. I'm not sure. It might be. It's a valley. An incredible valley. It's like the Auvergne. It's like an English Auvergne. Come on, sweetheart. You've got to have a look. It might be all right. It might be. I'm not sure. You've got to come though. Please come.'

She was in Sussex, having gone to look at a house that we both knew was too small – it was a converted observatory – and in the wrong place, on the top of a hill where nothing would ever grow, even if its views were to the Downs and the sea and a vast dome of observable sky. It was her second visit. She asked the man living there where he usually went for a walk. He mentioned a lane that dropped from this observatory down through the woods to the valley of a little river.

She had gone down the lane, curling between the hedgerows, under the branches of the overhanging trees. It was springtime and the anemones were starry in the wood. Primroses were tucked into the shade of the hedge banks. Catkins hung off the hazels over the lane. And then, at a corner, where to one side the trees opened out to a view down the valley, and from where the pleats of the valley sides folded in one after another into the blue distance of other woods and other farms, 4 or 5 miles away, there was a sign hanging out into the lane: 'FOR SALE'.

She took me back there the next day. Slowly the car went

down the lane. The flowers in the verges, the sunlight in blobs and patches on the surface of the road. The knitted detail of this wood-and-field place. If anywhere were ever to look like nurture, privacy, withdrawal, sustenance, love, permanence and embeddedness, this was it. Sarah had found it.

Even then, at the first instinct-driven look, this felt as if it might be the place. Why was that? And how can the mind, in a series of fugitive impressions, never analysed and perhaps not consciously registered, make its mind up so fast?

First, maybe, it was because Perch Hill was hidden. It was neither exposed to the world nor making a display to it. It was clearly living in its own nest of field and wood, a refuge which could find and supply richness from inside its own boundaries.

Second, I think, it reminded me a little of all the ingredients of the landscape I had known as a boy at Sissinghurst, 15 miles away in Kent: the coppice woods and the slightly rough pastures, the streams cutting down into the clay underbase, the woodland flora along their banks, that deep sensuous structure of light and shadow in a wooded country, where as you drop down a lane you are blinded first by the dazzle of the light and then by the depth of the shade, a flickering mobility in the world around you. Even at that subliminal level, here was somewhere that promised complexity and richness, secrets to be searched for and found.

And third, it was just the time of year, the first part of May, when England looks as if it has been newly made and the stitchwort and campion are sparkling in the lane banks and not a single leaf on a single tree has yet gone leathery or dark or lost that bright, edible, salad greenness with which leaves first emerge into the world; and when even though the sun is shining the air is still cold and you can feel the fingers of the wind making its way between your shirt and your skin, a sensation somewhere on the boundary of uncomfortable and perfect, as if nearly perfect, as if courting perfection.

I can make this analysis only now. Twenty years ago, we were driving blind.

We turned the corner, saw the agent's board, the sign on a little brick building saying 'Perch Hill Farm', and drove in. Almost everything about the place was as bad, in our eyes, as you could imagine it to be. The buildings were a horrible mixture of the improved and the wrecked: yards and yards of concrete; a plastic corrugated roof to the disintegrating barn; an oast-house whose upper storey had been removed during the war; a 1980s extension to the farmhouse, in the style of a garage attempting to look like a granary, paid for, I later learned, by selling off the milk quota. The farmhouse itself was dark and dingy downstairs. In most of the ground-floor rooms I was unable to stand up. Upstairs there was grey cheap carpet, gilded light fittings, down-lighters and pine-louvred cupboards. A 1940s brick cow shed had been enlarged with an extension made of telegraph poles and more corrugated sheeting. Three other sheds – for calves, logs and rubbish, I was told – lay scattered around the site looking as if they were waiting to be tidied up. Various bits of grass were carefully mown. There was a decorative fish pond the size of a dining-room table in front of the granary-sitting room. The truncated oast-house had become a cart shed but was now in use as an art gallery. There were places for customers to park.

None of this was quite what had been imagined. The smiles remained hanging on our lips. The buildings were raw-edged. Their arrangement was not quite what you would have hoped for, not quite a clustered yard, but a little strung out along the hill. The geese by the farm pond were angry. And a wind blew from the west. *Turn aside, turn aside. I have seen the sun break through* . . . The Bright Field murmurings were no more than faint.

Was it that time we walked around the farm or another? I

don't quite remember. We left again, slowly, back up the Dudwell lane, along others. We had lunch on the grass outside the Ash Tree at Ashburnham. I drank a pint of Harvey's bitter and the bees hummed. We were not sure. We went back on other days, again and yet again, taking friends with us. They all thought not. The place was trammelled. Whatever it might once have had was now gone. We heard somehow that the actor and comedian John Wells had looked at the place and rejected it: too much to do. We too should look elsewhere. So we did: a large fruit farm near Canterbury, other places, Brown Oak Farm, Burned Oak Farm, Five Oaks Farm, which I occasionally pass in the car nowadays, now the focus of other lives, diverged from ours like atoms that collide for an instant and then bounce on to other paths, and never to connect with ours again.

The Perch Hill valley would not go away. It had taken up residence in my mind. I bought the largest-scale map of it that I could find and kept it on my desk. I read it at night before going to sleep, walking the dream place: the extraordinary absence of roads, the isolated farms down at the end of long tracks, the lobes of wood and fingers of meadow, the streams incised into creases in the contours, the enclosed world away from the brutalizing openness which I felt had reduced me to the condition I was now in. It is a hungry business, map-reading. It only feeds the appetite for the real. I had drawn in red biro a line around the fields and woods that went with Perch Hill Farm: 90 acres in all, draped across a shoulder of hill that ran down to the valley of the River Dudwell, and entirely surrounded by the remains of Dallington Forest. The red line around these acres made an island of significance. The more I looked at the map the more real my possession of those fields became, the more that red biro line described the island reality for which we were both longing. 'Let's go again,' I said to Sarah. It would be the last time, the last throw, and then I would push this map away and the place would mean nothing to us and we could move on to other places and other obsessions.

It was a summer evening, four months after Sarah had first wandered down the lane. We went not to the house but to the fields. We had brought some bread and cheese with us. We walked around and the light was pouring honey on the woods. At the end we lay down in the big hay meadow known as the Way Field and looked across the valley to the net of hedgy woods and pastures beyond it, the terracotta tile-hung farmhouses pimpled among them, the air of unfiddled-with completeness, the haze of the hay. Owls hooted; two deer and their fawns came out of the wood into the bottom of the field to graze and look, graze and look in the pausing, anxious way they do. Graze and look, graze and look: it was what we had been doing for too long. We decided there and then: for the sort of money that could have bought you an extremely nice house in west London (double-fronted, courteous neighbours, Rosemary Vereyfied garden, a frieze of parked German cars, chocolate-coloured Labradors in pairs on red leashes) we would buy a cramped, dark old farmhouse, a collection of decrepit outbuildings and some fields that would never in a thousand years produce any income worth having. Was this wise? Yes. This was wise, the right thing to do, plumping for the lit bush. What else could money be for?

John Ventnor, the art dealer who owned the place, wanted what seemed like an outrageous amount for it: £480,000. We could afford, we thought, after the endless shuffling of portfolios, the sale of heirlooms and the accommodating of 'certain grave reservations' of financial advisers, no more than £375,000 and that was what we offered. Not enough. We offered £20,000 more. Not enough. What would be enough? He was prepared to countenance a 10 per cent discount on the asking price: £432,000. Too much. What about £410,000? Not enough. And there it stuck for months.

We began again to look at other places but none was right.

The vision in the evening field had its hooks in us. We waited, hoping that the delay would get to work on him, but it didn't. It became clear that he shared the freehold with a stepdaughter who no longer lived there. She wanted him to sell up but he didn't want to leave. He had no incentive to lower his price any further. We were in an impasse.

Sarah sold her house in London, our daughter Rosie was born and we all moved together into a rented basement flat of profound sterility and gloominess. I got a job on a newspaper and we borrowed money on that rather slender foundation. On winter evenings I drew plans and projections on my computer of how we might change Perch Hill, what kind of garden we might make, how we might take the land and farm it in a way that would be more generous towards it. The sliced-off oast-house became whole again on my night-time screen. I took to sitting in front of it with no other light in the room, a silvery brightness emanating from the dark, a possible future set against a present reality. Plantings and vistas criss-crossed the spaces between the buildings on my computer plans. One after another of these schemes I drew up, ever more elaborate, and they all shared the same title: 'Arcadia for £432,000.' *Give everything you have.*

One winter day I went down there again. I had never seen Perch Hill outside its springtime freshness or its hay-encompassed summer glory. This day was different. I was alone; Sarah remained with Rosie in London. A wind was cutting in from the east and for days southern England had remained below zero. All colour had drained out of the landscape. In the valley, the woods were black and the fields a silky grey as in my night-time visions of the place. The stones on the track into the farm were frozen and they made no sound as I drove over them. The geese were huddled in the lee of a bank, fingers of wind lifting the feathers on their backs. Even they couldn't bring themselves to run out and attack me. Frost-filled gusts blew across the frozen pond. The buildings were besieged by cold.

I was there to persuade Ventnor that he should sell his farm to us for something less than the £432,000 at which he had stuck for so long. I had no real tools or levers with which to achieve this, only the suggestion that a lower price might be fair. Smilingly, over a cup of coffee, he refused. We sat first in the kitchen and then by the fire. He was polite but adamant. The oak logs burned slowly. To my own surprise, I felt no resentment. I sat there agreeing with every word he said. Why should he leave the embrace of this? Why should the poor man go out into the cold if he did not want to?

He left the room to see a man who had come to the door and as I sat there I began to be embraced by the warmth of the house. I felt it wrap its own fingers around me. If a house could speak, that was the day it spoke, the day I learned this wasn't simply a place where we could come and impose our preconceptions. We couldn't simply land the Bright Field fantasy here and take that as the reality in which we were now ensconced. There was some kind of dignity of place to be respected here. It had a self-sufficiency which went beyond the demands and obsessions of its current occupants. There was a pattern to it, a private rhythm, the deep, slow music to which it had been moving for the four or five centuries that people had lived in it. The two of us men sitting here now in front of the fire, what were we in the light of that? Transient parasites.

I left and we had failed to agree on price but in some other, quite unstated way, I had succumbed. The buildings might be a mishmash of what we wanted and what we didn't. They might confront us with a list of things to do that stretched 10, 20 years into the future. The price that was required might be unreasonably high. But all of that was translated that frosty day, or perhaps started to be translated that frosty day, with the hot oak fire glowing as the only point of colour in a colourless world, into something quite different: a commitment to the place as it actually was, with all the wrinkles of its history and its habits, all its failings and imperfections, all its human muddle. Stop fussing, it said. Give yourself over to what seems good. Here – after catastrophe and culpable failure in my own life, after I had witnessed Sarah, now my wife, tending to me as I collapsed – was some kind of signpost towards coherence. Don't look for the perfect; don't be dissatisfied if the reality does not match the vision. Don't insist on your own way. Feed yourself into patterns that others have made and draw your sustenance from them. Accept the other.

'You mean pay him what he wants?' Sarah asked that evening as I put this to her.

'Yes,' I said. And we did.

If a son or daughter of mine said to me nowadays that they were thinking of doing what we did then, selling everything, taking on deep debts, putting their families on the verge of penury for years to come and acquiring a place that needed more sorting out than they knew how to pay for, a rambling collection of half-coherent buildings and raggedy fields, I would say, 'Are you sure? Are you sure you want to shackle yourself with all this? Do you know what it is you are so hungry for that this seems a price worth paying?'

Just now, faintly, as ghosts from the past, I remember people saying those things to us at the time and thinking, 'Ah, so they don't understand either. They haven't understood what it is to be really and properly alive.' And knowing for sure what that meant myself: that when faced with a steep slope or a rough sea, you should not quiver on the brink, or spend your life pacing up and down on the sand looking at the surf. You should plunge off down it or into it, trusting that when you arrive at the bottom or the far shore, you will know at least that the world's terrors are not quite as terrifying as they sometimes seem.

Do I believe that now? Nelson told his captains that the boldest moves were the safest, but Nelson was happy to lose everyone and everything in pursuit of victory. I can't forget the decades of debt and anxiety which lay ahead of us then, the years of work in trying to reduce the mountain of borrowings, article after article, the alarm clocks in the dark, the working late on into the night to try to balance the books. But would I now exchange the life we have had for one in which we had never taken that risk or made that step? No, not at all. I am as happy that Sarah and I married ourselves to Perch Hill as I am about the existence and beauty of my own children.

Green Fading into Blue

A CLOUD was down over the hill and the air was damp like a cloth that had just been wrung out. The buildings came like tankers out of the mist. Had we made a mistake? 'Is it a sea fog?' Sarah asked Ventnor.

'Oh no,' he said languidly, 'it's always like this here.'

Somehow his grief smeared us. He was unshaven; he had been unable to find a razor after he had packed everything. His mind was moving from one thing to another. This and that he talked about, these keys again, the oil delivery again, his own untidied odds and ends, a sort of humility in front of us as 'the owners' which grated as it reached us, as it must have grated as he said it. His eyes had black rings under them, wide panda-zones of unhappiness. Anyone, I suppose, would have been grieving at the loss of this place. It looked like an amputation. Even so, I felt nothing but impatience, as though it were already ours and he no more than an interloper here. He said nothing about that. What a curious business, this buying and selling of the things we love. It's like a slave trade. *Go, go,* I said to him in silence.

His mother-in-law was there with him. She was less restrained. She showed us pictures of their dogs cavorting in the wood. I felt like saying 'our wood'. She was still possessive. 'I'd hate to think of anyone making a mess in there after what I've done,' she said. I could see her primping the back of her hair and looking at me as though I were a piece of dog mess myself. And I suppose I was, in their eyes, the agent of eviction. *Go on, away with you.*

I was edged by it all. The house seemed ugly, stark and poky. I hardly fitted through a single door. Would it ever be redeemable? I was still standing off, waiting for the mooring line, but Sarah was sublime, confident, already arrived. 'Why do starlings look so greasy?' I heard myself asking Ventnor. 'Like a head of hair that hasn't been washed for weeks. They look like bookies.' He went at last, his sadness bottled up inside the great length of his long, thin body.

We waited for the furniture van. The house seemed inadequate for our lives. I picked some flowers, I looked at the view from the top field, our summit, and we waited and waited for the van. At last they called, about midday. They were in Brightling, lost. Sarah went to guide them in, while Rosie slept upstairs. The van came. It was too big to fit around the corner of the track past the oast-house and so everything had to be carried from the other side, 100 yards further. All afternoon our possessions rolled out and into the buildings, this clothing of the bones. *Come on, faster, faster.* The place started to become ours. It was as though the house were trying on new clothes. Sarah was worried by the sight of a staked lilac. Was the wind really that bad?

The removal men went. The oast looked like a jumble sale and the various rooms of the house half OK with our furniture. 'Change that window, pull down that extension, put the cowl of the oast back.' I could have spent £100,000 here that day. Sarah and Rosie went off shopping. Ventnor returned to collect a few more things. I didn't want to have to deal with him again. 'I see they've done some damage there,' he said, pointing at the place where the lorry's wheels had cut into the turf, trying to get around the sharp corner by the oast-house. I hadn't even noticed. I listened to his engine as he drove off towards somewhere else in England, the gears changing uphill to Brightling

Needle, and then down more easily the far side, the sound, like a boat's wake, slowly folding back into the trees. We never saw him again.

He had gone, it was quiet and I was alone for the first time in Perch Hill. I could feel the silence between my fingertips, the extraordinary substance of this new place, this new-old place, new-bought but ancient, ours and not ours, seeping and creeping around me. It was as though I had learned sub-aqua and for the first time had lowered myself gingerly into the body of the pool, to sense this new dimension thickly present around me as somewhere in which life might be lived and movements made. Until now all I had seen was the surface of the water, its tremors and eddies. Now, like a pike, I could hang inside it. I could feel the water starting to flow and ripple between my fingers.

That evening, as the sun dropped into the wood, I walked the boundaries, the shores of the island, the places where the woodland trees reached their arms out over the pastures. Here and there, the bluebells crept out into the grass like a painted shadow. Wild garlic was growing at the bottom of the Slip Field. I lay down in the Way Field, the field where we had decided to come here all those months before, a place and a decision which were now seared into my life like a brand, and as I lay there felt the earth under my back, its deep solidity, as Richard Jefferies had done 100 years before on the Wiltshire Downs. The hand of the rock itself was holding me up, presenting me to the sky, my body and self moulded to the contours and matched to the irreducibility of this hill on this farm at this moment. 'You cannot fall through a field,' I said aloud.

I took stock. What was this place to which we were now wedded? It had cost £432,000, plus all the fees. We had borrowed £160,000 to make that up, on top of the sale of the London house. My father was lending us £25,000. Our position was strung out and I had ricked my back. I felt as weak and as

31

impotent as at any time in my life. It was an intuitive under-standing, an act of faith, that the deep substance of this little fragment of landscape would mend that lack and make me whole. It was a farm of 90 acres in the Sussex Weald, about two hours south of London in the usual traffic, but no more than an hour from Westminster Bridge at five on a summer's morning. It was down a little lane, as obscure as one could feel in the south of England, with no sound of traffic, no hint of a sodium glare in the sky at night and an air of enclosure and privacy. At the edge of the land, by the lane, was the farmyard, with its utterly compromised mixture of bad old and bad new.

Beyond the farmyard itself, things improved. It was indeed one Bright Field after another. The buildings were at almost the highest point and in all directions the land fell away in pleats, like the folds of a cloth as it drops from the table to the floor. The creases were filled with little strips of wood under whose branches small trickly streams made their way towards the River Dudwell. The broad rounded backs of the pleats were the fields themselves, eleven of them, little hedged enclosures. They made up the small island block entirely surrounded by wood. That wood was part of the ancient forest of the Weald, whose name itself, cognate with *wald*, meant forest. The farmland was cut from it perhaps in the 15th or 16th century. The house was built in about 1580 and was probably made of the oaks that once grew where it stood. It was poor land, solid clay, high and windy all year, cold, wet and clammy in the winter, hard, heavy and cloddy in the summer.

Because nothing destroys a landscape like money, its poverty had preserved it. We were on the edge of viable agriculture here, one of the last pieces to be cleared from the wood and already in part going back to it. You could see the lines of old hedges, hornbeam and hawthorn, growing as 40-foot trees in the middle of the woods. Those were the boundaries of the ghost fields, abandoned and returning to their natural condition. Because it was so poor, it had never been worth a farmer's while to drive the land hard. That's why these fields are as beautiful as any you could find in Europe, or the world come to that.

Of course there are many other pockets of beauty in England, at least away from the great slabs of denuded arable land in East Anglia and the Midlands. In my 20s I had walked through many of them, about 3,000 miles on foot when writing a book about English paths, and then as a travel writer for the *Sunday Times* I had walked a great deal more. In the western counties, from Devon and Dorset, up the Welsh marches to Lancashire and Cumbria, I had fallen in love with a country I hadn't known until then, as a knitted thing, a visible testament to the long and intimate encounter between England and the English. It is the national autobiography, written every day for 1,500 years, with more life buried in it than any of us will ever know, with little thought ever given to its overall effect and its language often obscure. Maybe that is what we found at Perch Hill, a miracle of retained memory, a depth of time, and the mute, ox-like certainty which comes with that, away from the zigzags of our own chaotic existence. Nature was part of it, not all. This was no wilderness. Nor, though, was it an exclusively man-made place, sheeny and slicked up. Sometimes now I wish Perch Hill – our lives – had happened elsewhere in England, somewhere smarter and sleeker, with an elegant trout stream or smooth chalky views, but Perch Hill, stumbled on by chance, in all its scruffiness and lack of polish, but with its promise of what we always used to call *echt*ness, an authentic, vital beauty that came up from the roots, was the right place for us. Human and natural met there in a rough old encounter and that was the world we needed.

There was a line from a poem of Tennyson's which, from time to time that year, bumped up into my conscious mind, and presumably lurked not far beneath for the rest of it. 'Green Sussex,' it said, 'fading into blue.' That was this farm in a phrase:

the green immediacy, the plunge for the valley, the stepped ridges of the Weald, blueing into the distance 10 and 12 miles away. This wasn't a little button of perfection, a cherry perched on a cake of the wrecked, but part of a larger world and as I lay in the Way Field that evening all I could think of was the feeling of extent that ran out from there across the lane, down into the field called Toyland, beyond that into the valley of woods running off to the west, to the river down there, the deer nosing in that wood and the sight I had that morning, as we were waiting for the van, of the fox running down through that field, on the wood edge, no more than the tip of its tail visible above the grasses, a dancing point like the tip of a conductor's baton . . .

I shall not forget that evening. The spring was going haywire around me. It was DNA bedlam, nature's opening day. The black-thorn was stark white in the dark and shady corners. The willows had turned eau-de-nil. Oaks were the odd springtime mixture of red and formaldehyde yellow, the colour of old flesh preserved in bottles. The wild cherries stood hard and white like pylons in the wood and the crab apples, lower, more crabbed in form, were in full pink flower – an incredible thing a whole tree of that, the most sweetly beautiful flower in England, dolloped and larded all over the branches of a wrinkled, half-decrepit tree. In places nothing was doing better than the nettles and the thistles, but in the wood, there were the wall-to-wall bluebells, pale, almost lilac in the Middle Shaw, that eyeshadow blue: in the half-lit green darkness of the wood, that incredible, glamorous, seductive haze of the bluebell's blue, a nightclub sheen in the low light, the sexiest colour in the English landscape, hazing my eyes, a pool of colour into which, if I could, I would have dived there and then.

There were deer on the top field. The light was catching the ridged knobbles of their spines. I drank it up: this bright sunshine, even late, the bars of it poking into the shadows of the wood; the comfort of the grass; the lane a continuous mass of wood anemones, cuckoo flowers, primroses; and one very creamy anemone up by the gate. Its colour looked to me like the top of the milk.

'These foam-bells on the hidden currents of being', Hugh MacDiarmid once called spring flowers, and that attitude, a slightly dismissive superiority, used to be mine too. Geology, the understructure, the creation of circumstances: those were the things that used to matter to me. I preferred the hard and stony parts of the north and west or the higher places in the Alps where, after the snow has gone, the crests and ridges are left as abused and brutalized as any frost-shattered quarry. Walking across those high, dry Alps, I have seen the whole world in every direction desert-like in its austerity, a bleakness beyond either ugliness or beauty, and thought that life could offer me nothing more.

I still lusted after that, for all the clean hard-pressure rigour of that alienating landscape, serene precisely because it is so dreadful, because from that point of desiccation there was nowhere lower to fall. But alongside it now, there was this other thing, this undeniable life-spurt in the spring, whose toughness was subtler than the stone's but whose persistence would outlive it. Genes last longer than rocks. They slip through unbroken while continents collide and are consumed. These plants, I now saw, were the world's version of eternity, the lit bush. If you wanted to ally yourself with strength, nothing would be more sustaining than the spring flowers.

Over the following days, we dressed the house as though preparing it to go out. It was like dressing a father or mother. She sat there mute while coats and ribbons were tried on her. 'Oh you look nice like that,' we would say, 'or that, or maybe that.' Rooms acquired meaning, another meaning. In the kitchen, painted on the cupboard door, I found a coat of arms: six white feathers on a blue ground and the motto '*Labore et perseverantia*',

By work and perseverance. It looked fairly new. What was it, a kind of d'Urberville story of a noble family collapsed to the poverty of this, to the resolution of that motto against all the odds? It was certainly a failed farm. That was the only reason we were there. We had crept like hermit crabs into a shell that others had vacated.

But I guessed the arms and the motto had no great ancestry. Had Ventnor himself painted them here in the last few years? I knew that he had attempted to make a business out of this farm, to continue with the dairy herd that he had bought with it. But he didn't know what he was doing and had lost a fortune. The last thing he asked me before he left was 'Are you thinking of farming here?' to which I had been noncommittal. With something of a glittery eye, he told me not to consider it. That was a sure route to disaster. Soon stories were reaching me of Ventnor sitting in the kitchen here, his unpaid bills laid out on the table in front of him like a kind of Pelmanism from hell, his head in his hands, his prospects hopeless. Every one of these stories ended with the same warning: Don't do it.

The Ventnor experience seemed somehow to stand between us and an earlier past. He was one of us, an urban escapee, a pastoralist, who had turned the old oast-house into an art gallery and put coach-lamp-style lights outside the doors. Where was the contact with the real thing, the real past here? There was a glimmering of discontent in my mind about that gap, still a glass wall between us and the essential nature of Perch Hill. Before Ventnor, I knew, the last farmers here had been called Weekes. Where were they? Had all trace of them disappeared? It wasn't long before we realized that the very opposite was the case: Ken and Brenda Weekes still lived in the cottage 200 yards away across the fields. A day or two after our arrival, Sarah and I went up to see them and from that moment they became a fixed point in our lives.

The boys were here and were shrieking in the new expanse

of garden. Tom was ten, Will eight and Ben six. We planned bike routes across the fields and began making a tree house in the Middle Shaw, nailing and binding scaffold boards and half-rotten ladders to the ancient twisting hornbeams. We got a giant trampoline and put it in the barn, where the three of them competed with each other, trying to touch the tie-beams high above them. Sarah and I were anxious and buoyed up in equal measure at what we had taken on. Across the fields, we could hear Ken mowing the lawn around his cottage: the sound of a half-distant mower in early summer, a man in shirtsleeves and sleeveless jersey, his dog on the lawn beside him, the sun slipping in and out of bubbled clouds, and all around us, to east and west, Sussex stepping off into an inviting afternoon. It was, in a way, what we had come here for.

We walked up there, not across the fields that first time but up the lane. The hedge was in brilliant new leaf. Ken and Brenda came to their garden gate, asked us in, a cup of tea in the kitchen, Gemma the dog lying by the Rayburn, and a sort of inspecting openness in them both, the welcome mixed with 'Who are you? What sort of people are you?' I shall always remember two things Ken said. One with his tang of acid: 'You know what they always said about this farm, don't you? They always said this was the poorest farm in the parish.' The other with the warmth that can spread like butter around him: 'That's one thing that's lovely, children's voices down at the farm again. That's a sound we haven't had for a long time here.'

To a degree I didn't understand at the time, we had entered Ken Weekes's world. Perhaps we had bought the farm, perhaps the deeds were in our name, perhaps we were living in the farmhouse, perhaps I was meant to be deciding what should happen to the woods and fields, but none of that could alter the central fact: Perch Hill was Ken's in a deeper sense than any deed of conveyance could ever accomplish.

He had come here in 1942 as a six-year-old boy to live in the house we were now occupying. His father, 'Old Ron', was farm manager for a London entrepreneur and 'a gentleman, one of the real old gentry', Mr George Wilson-Fox. 'Old Wilson' used to come down with his friends on a Saturday. The Weekeses would all put on clean white dairymen's coats to show the proprietor and his guests the herd of prize pedigree Friesians, spotless animals, their tails washed twice a day every day, the cow shed whitewashed every year, a cow shed so clean 'you could eat your dinner off that floor'.

It was a place dedicated to excellence. Wilson-Fox made sure there was never any shortage of money for the farm and Ron imposed his discipline on it. 'The cattle always came first,' Ken remembered. 'Even if you were dying, you had to look after the cows. I remember Old Ron kicking us out of bed to go and milk the cows one morning when I could hardly move – "Come on, you bugger, get out, there's work to do" – and it was so cold in there in the cow shed with a north-easterly that the milk was freezing in the milk-line. But we got it done. It all had to be done by eight in the morning if you wanted to sit down to breakfast. You couldn't have breakfast unless the cattle had been looked after first.'

It is a lost world. Nothing like these small dairy farms exists here any more. They have all gone and Ken has witnessed their disappearance, the total evaporation of the world in which he grew up. About that he seems to feel bleak and accepting in equal measure. Every inch of this farm carries some memory or mark of Ken's life here: the day the doodlebug crashed in the wood at the bottom of the Way Field; those moments in Beech Meadow where Old Ron, in late June or early July, would pick a bunch of flowers for Dolly, Ken's mother, the signal for the boys that haymaking was about to begin; the day the earth suddenly slipped after they had ploughed it in the field for ever after known as the Slip Field. A farm is a farmer's autobiography and this one belongs to Ken.

When he married Brenda, in 1959, they moved into the cottage across the fields. His mother and brother stayed in the farmhouse. Wilson-Fox died in 1971, but Ken continued farming for the trustees of the estate for another 15 years until, in 1986, the farm was sold, along with its herd of cattle, to John Ventnor, who wanted to be a farmer. Ken stayed in the cottage, and set the art dealer on his way, helping him for a year or so, but they fell out. 'We had a misunderstanding,' Ken said. For several years after that Ken was not even allowed to walk his dog in these fields.

There was something of a false cheeriness in both of us as we talked. Each of us was guarded against the other. But the afternoon floated on Ken's stories. He could remember watching the pilots of the Luftwaffe Messerschmitts, low enough for you to see a figure in the cockpit. 'Oh yes, you could see them sitting in there all right.' One evening the Weekeses were all down in the Way Field getting in the hay and there were so many of the German planes that his father said they'd better go in. 'You could never tell, could you? Bastards.' Ken's performance culminated in his favourite story about the hunt. He was out in the Cottage Field, tending to one of the cows which was poorly, when he chanced to look up and see a whole crowd of the hunt come pouring down the trackway that leads off the bridlepath and into the Perch Hill farmyard.

'"I say,"' Ken bellowed at them – 'because they'll only under- stand you if you talk to them in a way they do understand – "why don't you fucking well bugger off out of there?" And,' Ken says, looking round, all smiles, 'do you know what? They did!'

Another piece soon dropped into place. Will Clark came up one day from the village. He had been doing some work on the farm for John Ventnor. Peter, his son, had been working in the wood and mowing the grass. Ventnor had said that there was no one he could recommend more highly. And that's how it turned out. As soon as Will walked into the yard I could see what he meant. His eyes were the colour of old jeans. He swept his hair in a repeated gesture up and over his forehead into a wide long curl that could only be the descendant of a rocker's quiff, 40 years on. He smiled with his eyes and talked with a laugh in his voice. 'We'll be haying soon,' he said. He cuddled Rosie. His taste in shirts was perfect, lime green and tangerine orange, unchanged, I guess, since he was bike-mad in the fifties, when he used to do a ton down the long straight stretch to Lewes called The Broyle, or burn up and down the High Street in the village to impress the girls. He was the only man in Burwash ever to get to Tunbridge Wells in 12½ minutes, or so he told me. He talked broad Sussex: fence posts were 'spiles', working in the mud got you 'all slubbed up', anything that needed doing always involved 'stirring it about a bit', a sickle was 'a swap'. He had started his working life when he was 14 on a farm at Hawkhurst, just over the border into Kent, looking after the horses. He knew all about machines and wood and wooding and how to get a big ash butt out of a difficult corner. He was the man of the place and he would be the man for us. Will had been ill for years – his kidneys scarcely worked and he had to spend three hours a day at home on a dialysis machine – and he said that Peter would have to do the heavy work. 'He's the muscle man.' And so we plugged in. This other world was closing over us, some version of pastoral folding us in its lap.

All the same, I was anxious about it all. The stupidity of what we were doing was brought home, involuntarily or not, when people we knew from London dropped in. There was always a vulture in the party, someone who would unerringly make you sour with a remark. 'Oh yes,' one of these people said in the early days, nosing around the ugliness of our horrible buildings, 'it's a very nice *spot*, isn't it?' A very nice spot: the silent pinchedness of what is not said. Why do these people wreak destruction? Why do they do such dishonest damage? I couldn't believe how soured I felt by them. But why should I have been? Why did

I even care? Perhaps because the whole point of Perch Hill was to take ourselves out of range of their criticisms, their worldly knowingness. Now, I am sure, nothing they said would come anywhere near me. It is one of the consolations of age that your own self-knowledge allows other people's criticisms to break around you like little waves. But then, in our tender state, to have their all-too-predictable strictures applied to our precious refuge was like an experience you occasionally get as a writer. You have written something which matters to you and which tries to say something beyond what is ordinarily said, and as a result is likely to be a little rough at the edges. Your reader looks at it, but they don't read into the heart of it, the point of it, and stay critically on the edge, looking at the punctuation or the length of sentences or, worst of all, the definition of terms. I once wrote a book about a place I loved and which, on its first page, mentioned the 'branched orchids' that grew there. A woman told me casually one day that she hadn't got any further than that first page because 'There aren't any branched orchids.' I have never been able to look at an orchid since without feeling with the ends of my fingers for those tiny branches on which each of its individual flowers sits.

Now, though, in retrospect, I get the point: Perch Hill is a nice spot but there was nothing nice about its buildings. The judgement was correct. But Sarah and I were not living in the world of correct judgements. And our visitors from London could never have understood the powerful psychic reality here: the way I wanted to wriggle under the skin of this place so that only my eyes were above the skin of the turf like a hippo in its river and the bed of green comfort around me, the osmotic relation to place so that there was no distinction between me and it, no boundary at the skin. Of course they couldn't, because that is not something that can be said in polite society. It was that kind of pre-rational understanding that I was after, like a dog rolling in muck.

We didn't know what we were doing. We arrived on this farm as naked as Adam and Eve and we were setting about making it right. We knew what we wanted – a sense of completeness. That sounds so vague now and perhaps it was. But there were real models in our minds. As a boy at Sissinghurst, I had known a kind of completeness in the world that surrounded me, a house and garden, farm, woods, streams and fields, with a sense of that pattern continuing beyond its boundaries in much the same way, to be explored on foot through the woods and hay meadows, by bike down the long sinuous lanes which only decades later did I realize were the drove roads by which the Weald was first settled. That was a memory which seemed to have all the elements of a life – adventure, energy, people, community, love, beauty.

And then, more recently, Sarah and I had stayed for a few months in a cottage next to a house belonging to John and Caryl Hubbard in Dorset, at Chilcombe, a tiny settlement with its own tiny church, looking out over a theatre of fields and woods that led down to the shingle bank at Chesil Beach. Here too house and garden and chickens and sheep and cattle and the whole wide view and the sense of Dorset and England – with, miraculously, the strip of shining sea laid above it all – were folded in together in a way that is simply not accessible in any city.

This is not an aberrational idea. It appears at the earliest moments of Europe, in Minoan Crete, in the Bronze Age 4,000 years ago, when the priest rulers of that civilization made for themselves small and elegant country houses, surrounded by flower gardens, vineyards, orchards and olives groves, in carefully and beautifully chosen places where the cultivated country and the distant mountains were laid out around them like the background to a Renaissance portrait. This is the vision of the Horatian farm, an easy place because it is at ease with its surroundings; it is the ideal behind the Palladian villas of the late 16th-century Veneto; it is the transformed vision in eighteenth-century England and the English seaboard of America; it lies behind the ideals of Ruskin and Morris, driven by a need for an intimacy with the natural which goes beyond the crude act of *buying* which is at the root of all cities. That is what completeness meant and means to me: an entirely full and committed engagement with the real world in all the dimensions which the world can offer.

We didn't quite know how to get there. All we could do was stumble off into the dark, hoping and trusting that our instincts were right. That was the point. The whole enterprise was a blunder into truth, wobbling chaotically towards the goal. It was good because it was messy. If it had all been neater, if we had known what we were doing, it wouldn't have had the juice in it. The whole thing would have flattened out in the drear of expertise. As it was, ignorance was the great enabler and incompetence the condition of life. Or so I would say to myself in my storming rage after the nay-sayers had gone.

Sometimes I felt we were surrounded by know-alls. Not the people who lived near us, the Will Clarks, the Ken Weekeses, who approached our efforts with a delicate sense of neither wanting to intrude nor wanting us to come too much of a cropper. No, the real killer know-alls were the partly ruralized urbanites who had acquired the cultural habit of telling other people what to do. It probably stemmed from the prefect system at public schools, compounded by middle-class careers in which the only necessary skill is the ability to disguise bossiness as brains.

You could see them heaving into view a mile away. They were struggling with their mission to inform. They knew they shouldn't. They felt they must. They wished they didn't have to. But they knew they ought to. One has a duty, after all. It's a responsibility to the landscape as a whole. And it would be so sad, wouldn't it, if it all came unstuck in the end for Adam and Sarah?

Out came the supercilious smiles. These were the opening, but doomed, attempts at a spirit of generosity. Soon enough they gave way to the barrage of assured, you-really-should-have-asked-me-first, pain-in-the-neck blather. The spirit hit the iceberg and sank.

No area of life was immune. I remember, classically, having our stack of firewood analysed by a man who, from what he was saying, was obviously chief firewood analyst for Deutsche Morgan Grenfell. Not much was right with the way we had done things. The shed was wrong; it needed more air holes, its roof was not very nice, the walls were unsatisfactory and it was in the wrong position. We shouldn't have bothered to put either chestnut or larch logs in the pile because they both spit when they burn and that wasn't good for the kiddies, was it? The oak was useless; it only burned on a massively hot fire, which we would never achieve because the rest of the wood was such rubbish. The ash had been split far too small and would burn too quickly. Sycamore had no calorific value to speak of and what we had was rotten. It would take more energy to start the fire than would be given out by it. And were we two years ahead with our cutting programme? He looked at me in that generous, hesitating way people use to those whose self-esteem they have just bulldozed into a silage pit.

The idea of putting up a building of any kind was a mistake. You would make the windows too small. You would spend too much money on it. You would do something totally out of character. You would create a dreadful ersatz fake ('Tesco's') when people in your position had a responsibility to patronize new architects and architecture. You were living in a retro hell. You would not install the correct insulation/safety features/ heating system. Heating systems! May I never, ever have another discussion about heating systems for the rest of my life.

Then there was the chicken question. You were thinking of getting far too many of them. Were you really going to be eating 80 eggs a week? Had you performed the cloacal swab test for salmonella on them yet? You certainly couldn't think of giving eggs away if they had dirty cloacas. Their housing was disgusting and if an RSPCA man should happen by, he would be appalled. You may have heard someone was prosecuted for just this kind of thing the other day. Anyway, they should have been bedded down on sawdust not straw. It was amazing you hadn't found that out for yourself. I don't quite understand why you were going in for these things.

Moving quickly on, children should wear clothes made out of only naturally occurring materials, fed only naturally occurring foods and baked beans should be sugar-free. Trees – these two subjects always somehow elide – should be planted without stakes, or tied only loosely to stakes, or planted without tree-guards or only after a comprehensive drainage system has been installed, or only on M25 rootstock, or only from Deacon's Nursery in the Isle of Wight, or only with local genetic material, gathered from the last of the local orchards, and anyway fruit trees are only a pleasure if you have done all the grafting and training yourself. Have you managed to do that, Adam? Or have you ever thought you might be taking on a little much here? Have you had your dog castrated yet? Aaaaaaaargh.

My sons – lovely stage in life – had just then started playing Oasis on their ghetto-blaster at crow-scarer volume. The songs wormed their way into the mind, colonizing whole stretches of it. After one particularly gruesome hour or two with a couple of people who came to lunch and knew every damn thing there was to know about the usual subjects – orchards, firewood, chickens, ducks ('one says "duck", doesn't one, in the plural?'), heating systems, woodland management, grants ('We've found we've done quite well out of the whole County Council Heritage Landscape scheme but I'm told, I'm afraid, that they've run out of money now and won't be taking any more applicants at least until fiscal 2000') – I found myself stacking the dishwasher and

45

singing, much too loudly, 'Yer gada roll wiv it/ya gorra take yer time/yer garra say wotya say/dern ledd anybuddy gerrin yer waaajy . . ./There's nuthin lef for me to saaiy . . .'

The bravado papered over a pit of anxiety. One morning I woke at four and said, aloud, 'I'm worried about the fields.'

'For Christ's sake,' Sarah said without a momentary flicker. She'd heard this sort of thing before and, anyway, was already awake worrying about the garden. We lay there in silence for a moment, travelling through the universe together at 24,000 miles an hour, each in a private little cubicle of hysteria and each thinking the other stupid. 'The fields are fine.'

'They aren't. What's wrong with the garden?'

'It's out of control.'

'That,' I told her, 'is also what's wrong with the fields.'

'I've never heard anything so ridiculous. Fields don't get out of control. That's one of the good things about them. They just sit there perfectly in control for day after day.' Gardens didn't, apparently. Gardens were nature on speed. In fact, you could see them as hyperactive fields. They went mad if you didn't look after them. Anyway the fields were not going to be photographed on Wednesday, were they?

This was true. Sarah had got a job as a junior doctor in the renal unit of the hospital in Brighton. She was sharing it with another young doctor but even half-time, in the days of famously long hours for junior doctors, it often required her to be there 40 hours a week. The hospital was about 45 minutes' drive each way. Day after day, she had to leave early and return late, missing Rosie, feeling that she had come here to find a new life with us but that her work was taking her away to the point where our lives were hardly shared at all. Even when at home, she was too tired to take much pleasure in what we were doing, or where we had come to. It seemed absurd.

We decided together that she couldn't go on. When she had

been pregnant with Rosie and after she was born, Sarah, who is incapable of doing nothing, had taken time off and set up a florist's business called Garlic & Sapphire with her university friend Lou Farman. As florists do, they had bought their flowers and foliage from wholesale merchants in Covent Garden market. This was fine, but it was not that easy to make any kind of living, and anyway it felt a little tertiary: selling flower arrangements to London clients from boxes of flowers bought at a market and almost entirely shipped in from industrial-scale producers in Holland. We could do better than that.

Sarah's father, a classics don at King's College, Cambridge, had also been a passionate botanist who with his own father had painted the entire British flora and was the co-author of the New Naturalist volume on mountain flowers. He had taken Sarah as a girl botanizing across bog, heath, mountain, meadow and moor in England, Scotland, Italy and Greece. She had drunk flowers in at his knee and from him had learned the science of flower reproduction and habitat. The Ravens had also made inspirationally beautiful gardens at their house near Cambridge (chalky) and around their holiday house on the west coast of Scotland (acid). So this much was clear: Sarah had gardens in her genes. She had to make a garden. Her life would not be complete unless she did.

We had made together a small and lovely garden in London with pebble paths and hazel hurdles and we had talked together about making it more productive. One day I happened to be sitting next to the publisher Frances Lincoln at a wedding party. 'Do you know what you should publish?' I said. 'No,' she said, a little wary. 'A book called *The Cutting Garden* about growing flowers to be cut and brought inside the house.' 'Are you going to do it?' she asked. No, I wasn't, but I knew someone who could.

So this was already in our mind when we came to Perch Hill and it was obvious that when Sarah gave up her job as a doctor, to look after Rosie and to be with us at home, she should embark on making the cutting garden and writing the book. We wrote the proposal together, with plans, plant lists and seasonal successions, and sent it off to Frances, and soon enough it was commissioned. Perch Hill was about to take its first step to new productivity.

It was to be a highly and beautifully illustrated book whose working title at home anyway was *The Expensive Garden*. From time to time in various parts of the house I used to find half-scribbled lists on the back of invoices from garden centres spread across the south of England, working out exactly how much had been spent on dahlia tubers, brick paths, taking up the brick paths because they were in the wrong place, the new, correctly aligned brick paths, the hypocaust system for the first green-house, the automatically opening vent system for the second, the underground electric wiring for the heated cold frames (yes, heated cold frames), the woven hazel fencing to give the correct cheap, rustic cottage look (gratifyingly more expensive than any other garden fencing currently on the market) and the extra pyramid box trees needed before Wednesday.

The consignment of topiary was delivered, one day early that summer, by an articulated Volvo turbo-cooled truck, whose body stretched 80 feet down the lane – it had caused a slight traffic rumpus on the main road just outside Burwash, attempting to manoeuvre itself like a suppository into the entrance of the lane – and whose driver with a flourish drew aside the long curtain that ran the length of its flank, saying 'There you are, instant gardening!' He must have done it before.

The inside of the lorry was a sort of tableau illustrating 'The Riches of Flora'. It contained enough topiary to re-equip the Villa Lante. Species ceanothus, or whatever they were, sported them-selves decorously among the aluminium stanchions of the lorry. The rear section was the kind of over-elaborate rose and clematis love-seat-cum-gazebo you sometimes see on stage in *As You Like It*. Transferred to the perfectly unpretentious vegetable patch maintained by the Weekeses, the disgorged innards of the Volvo turned Perch Hill Farm, instantly, as the man said, into the sort of embowered house-and-garden most people might labour for 20 years to produce. A visitor the following week congratulated Sarah on what he called 'its marvellous, patinated effect'. Some patina, I said, some cheque book.

At that stage, the advance on *The Expensive Garden* had covered about fifteen per cent of the money spent on making it. If even a tenth of that amount had been spent on the farm we would already be one of the showpieces of southern England. 'That point,' Sarah was in the habit of saying, for reasons I have never yet got her to explain, 'which you always make when we have people to supper, is totally inadmissible.'

But I was serious about the fields. I wanted the fields, which were beautiful in the large scale, to be perfect in detail too. I wanted to walk about in them and think, 'Yes, this is right, this is how things should be. This is complete.' That is not what I was thinking that summer. In fact, the more I got to know them, the more dissatisfied I became. Hence the 4 a.m. anxieties. The thistles were terrible. Some fields were so thick with thistles that my dear dog, the slightly fearful and profoundly loving Colonel Custard, refused to come for a walk in them. He stopped at the gate, sat down and put on the face which all dog-owners will recognize: 'Me through that?'

In the early hours, I used to have a sort of internal debate about the fields. It came from an unresolved conflict in my own mind, which could be reduced to this question: was the farm a vastly enlarged garden or was it part of the natural world which happened to be ours for the time being? The idea that it might be a business which could earn us money had never been seriously entertained. We might choose to have sheep, cows, chickens, ducks and pigs wandering about on it, but only in the same way that Sarah might order another five mature tulip trees for a little quincunx she had in mind. I would have animals because they looked nice.

To the question of big garden or slice of nature, I veered between one answer and then the other. In part it *was* like a huge, low-intensity garden. We were here because it was heart-stoppingly beautiful and one of the things that made it beautiful was the interfolding of wood, hedge and field. If the distinctions between them became blurred then a great deal of the beauty would go. The fields must look like fields, shorn, bright and clean, and the woods must look like woods, fluffy, full and dense. Field and wood were, here anyway, the rice and curry of land-scape aesthetics. Scurfy fields, as spotty as a week's stubble on an unshaven chin, looked horrible, untended, a room in a mess.

There were other things to think of too. If we simply mowed the fields to keep them bright and green or, horror of horrors, sprayed off the docks and thistles, we would not be attending to other aspects of the grassland which of course are valuable in themselves. There were clumps of dyer's greenweed here, whole spreads of the vetch here, called eggs and bacon and early purple orchids on the edges of the wood. Sprays would wreck all that and you had to allow those things to set seed and repro-duce if they were going to continue. You had to allow them to look messy. What to do? Obviously I had to learn how to manage grassland properly. I was blundering around in my ignorance.

I had a meeting one Saturday morning with the Wrenns, Brian and Stephen, father and son, our farming neighbours from Perryman's, on the other side of the hill. We were all a little shy with each other in the kitchen, too ready to agree with what the other had said.

Brian, sliding the conversation sideways, told me there were nightingales in our woods and nightjars and wheatears. I knew nothing about these things. We talked about the way that all the farmhouses here face south, their fronts to the warmth, their backs to the wind. How ingeniously the first people to settle here spied out the land. Stephen talked about the poverty of the

Weald, the way there is no topsoil, all the fertility poured away down the streams to the southern rich belt, the champaign country of rich southern Sussex. 'This is the poorest, the last bit of ground to be taken,' he said, 'but that's what has saved it.'

Then I said, with the Nescafé in me, we should talk about the real matter in hand. Brian turned businesslike. 'We would certainly like to have all the grass. But it's too late this year to get any nitrogen on for the silage. We'll get some heifers on to graze it later.'

'Of course,' I said, feeling I was about to introduce some urban gaucherie, 'and anyway, from our side, we would like to manage it in as much of a conservationist way as we can.' How I hate that word, but to my amazement they lay in happily with it. Even the air between us became somehow emotional at the recognition of shared ideas. Stephen talked about planting some of Perryman's with willows to provide fuel for a wood-burning power plant. That toboggan feeling developed around the table that we were not such aliens to each other as we might have imagined. The future opened like the curtains of a theatre. We all came to an agreement: they should take some of the grass that year as summer grazing. They would pay £1,000 for it, which was better than nothing I supposed, about 1.5 per cent return on capital. The real question was: were we prepared to forgo the £3,000-odd we would get from a conventional let for the sake of it looking nice and it being lovely? For saying no to nitrogen and no to high-pressure farming? Were we rich enough for that?

With a new bill-hook, I cut hazels in the wood for Sarah's new cutting patch. The old hazel was growing in stools 8 feet across. The middle of them was a jumble of old fallen sticks. One stick had rotted entirely on the inside but the skin had remained whole as an upright paper tube. There were deer in the wood, their awkward big bodies breaking the trees they hurried past, and around my feet tall purple orchids. I was cutting with slashing blunt incompetence at the hazels, half-tearing them away, but I loved their long swinging creak next to my ear as I carried them home on my shoulder. I felt the sweat run down my side in single, finger-sized trickles and I loved the smell, the woody, sweet, bruised, tannic vegetable reeking of the cut wood. That's what I was here for, the under-sense, those deeper connections, that core of intimacy.

I demolished the fence around the pond, shirt off, sweat and exhaustion. I sold the tractor on Will's advice and we were for the moment tractorless. Will was looking for another more reliable one. I made teepees out of hazels from the wood and some hurdles out of split chestnut. I tried to buy a mower but my credit card failed. It was the usual humiliation in front of a queue of men who were more interested in that little human drama than anything else. Rachel the shop girl blushed. I laughed it off. I needed to earn more money and I had a feeling we were veering, slowly but quite deliberately, towards a financial crisis.

But I loved it here that first spring. I loved the sustenance of the green, the kestrel that came daily and hung above the corners of the barns, moving from station to station. I loved the little seedlings of the oak and aspen sprouting in the grass along the wood edges. I loved the knit of the country, the jersey of it. I loved the sight of the ducks, two wild mallards on the pond, I loved the substance of the place, my new fax machine, the gentleness of Will and Peter, Rosie playing in the garden with her new nanny, Anna Cheney, who only years later would dare tell us exactly how horrified she had been at the chaos in which we were living, but how one thing had convinced her to come: the sight of Rosie's face as she sat at the kitchen table, so round and so sweet.

The garden accumulated. I tended to the house as I had tended no house before, tidying and hoovering. Sarah and I both had the feeling, if we were honest, that we should have waited to buy somewhere with more beautiful buildings but there we were: we couldn't say that now. This would be lovely in the slow unnoticed growth of the place around us. Forty years later, as we died, we would look at it and say, 'That was beautiful.' Life would be over, having been lived. The moments of revelation are all there is. This is all there is. This will be the undernote of my life: the making with a purpose, not the drifting of the survivor. Make, and you will be happy.

Part Two

MENDING

The Darting of Life

THAT SUMMER of 1994 burned. The south of England was bleak
with heat. Cars along the lane raised a floury dust in their wake.
The cow parsley and the trees in the hedges were coated with
it like loaves in a bakery. The streams were dry coming off the
hill and the river in the trench of the valley was little more than
a gravel bed across which a line of damp had been drawn,
connecting the shrunken pools.

I spent long days down there in the dark, deep shade of the
riverside trees. The valley felt enclosed, a place apart, and secrecy
gathered inside it. Rudyard Kipling lived here for the second
half of his life – he bought Bateman's, a large 17th-century iron-
master's house just below the last of our fields, in 1902 – and
the whole place remained haunted by his memory. Everywhere
you went, he had already described. It was here, among the
hidden constrictions of the valley where, in Kipling's wonderful
phrase, 'wind prowling through woods sounds like exciting
things going to happen', that I felt most in touch with where I
had come to live. This was the womb.

It was a pathless place, or at least the only paths were the old
deeply entrenched roads, never surfaced, which dropped from
the ridge to the south, crossed the river at gravelly fords and
then climbed through woods again to the ridge on the other
side. They were the only intrusion in what felt like an abandoned
world. The woods were named – Ware's Wood, Hook Wood,

Limekiln Wood, Stonehole Wood, Great Wood, Green Wood –
but it felt as if no one had been here for half a century. Hornbeam,
chestnut, ash and even oak had all been coppiced in the past
but none had been touched for decades. The marks of the great
combing of the 1987 storm were still there: 80-foot-tall ash trees
had fallen across the river from one bank to the other. The ivy
that once climbed up them now hung in Amazonian curtains
from the horizontal trees. Growing from the fallen trunks, small
linear woods of young ashes now pushed up towards the light.

I stumbled about in here, looking for some kind of inaccessible essence of the place. The deer had broken paths through
the undergrowth. The clay was scrabbled away where they had
jumped the little side streams. The fields of underwood garlic
had turned lemon yellow in the shade. And through it all the
river wound, curling back on itself, cutting out promontories
and peninsulas in the wooded banks, reaching down to the
underlying layer of dark, ribbed, iron-rich sandstone. Where it
cut into an iron vein, the metal bled into the stream and the
water flowed past it an almost marigold orange. This too was
Kipling's world, virtually unchanged since he had described it,
90 years before, in *Puck of Pook's Hill.*

> Even on the shaded water the air was hot and heavy with
> drowsy scents, while outside, through breaks in the trees,
> the sunshine burned the pasture like fire . . .
> The trees closing overhead made long tunnels through
> which the sunshine worked in blobs and patches. Down in
> the tunnels were bars of sand and gravel, old roots and
> trunks covered with moss or painted red by the irony water;
> clumps of fern and thirsty shy flowers who could not live
> away from moisture and shade.

As you pushed up through this wooded, private notch in Sussex,
so many miles away from the bungalowed, signposted and

estate-agented ridge-top roads, the river shrank still further to inch-deep pools and foot-wide rapids where banks of gravel had dammed the flow. In that clear, shallow water, life was exploding. Midges in Brownian motion were flashing on and off in the rods of sunlight that were rammed through the trees. Across the lazily moving water, insects drifted as slowly as those half-transparent specks that float across the surface of your eye. Far below, an inch away on the floor of the stream, their shadows tracked them, dark, four-petalled flowers easing across the stones. The dog snapped his chops at the passing bronze-backed flies. A 3-inch-long worm, as thin and as white as a cotton thread, snaked through the water and then under a stone. The stream was full of little shrimps, lying immobile on the gravel or wriggling there like rugby players caught in a tackle or, best of all, jetting around above it as fluidly as spaceships in their fluid medium, the darting of life.

By the middle of July, in all our hayfields, the grass was crisp and it was already late for the haying. That's what they call it here. Not haymaking or the hay harvest, but straightforward 'haying' on the same principle as lambing or wooding. It's the climax of the grass year and, as nothing except grass and thistles will grow on this farm, these few days became the point around which everything else revolved. It was high summer. Even as it was happening you could feel the winter nostalgia for it. You won't get a cattle or sheep farmer to talk starry-eyed about haymaking, but there was no doubt, in one sense anyway, that was how they felt. 'Look at that,' one of my neighbours would say to me during the following winter about a bale of his own hay he was trying to sell me. 'You can smell the summer sunshine in it,' and he buried his nose in the bale like a wine-taster in the heady, open mouth of his glass.

A friend rang up from London as we were about to start. 'Make hay while the sun shines,' he said to me on the phone, as if there were something original about the phrase. But it hardly needed to be said. Anxiety hovered over the beautiful fields.

They were beautiful. The buttercups and the red tips of the sorrel gave a colour-wash to the uncut grasses, a shifting chromatic shimmer to the browning fields. The enormous old hedges had thickened into little, banky woods so that the hay, even though it wasn't very thick that year, was cupped in their dark green bowls, a pale soup lapping at the brim.

Ken Weekes had been trying to persuade me all year that what was needed was a good dose of chemicals for the thistles and a ton or two of nitrogen to make the grass grow. We were already squabbling like a pair of old spinsters and I was relying on him for everything I did. Ken told me I needed Fred Groombridge, the sheep man from the village. Fred came down. He looked at me with only one eye, as though permanently squinting at the sun. 'That's because he's thinking with the other one,' Ken said.

I sold most of the hay to Fred as standing grass, £17.50 an acre. I had asked for £30, Fred suggested £15. He budged an inch, I moved a mile, but in return for that absurdly low price, Fred would also cut, turn, row up and bale 6 acres for me, 500 bales, which I would then have carted into the barn at my own expense. 'Perfect,' Fred said with his left eye, grinning. 'It's only money.'

Fred brought down his wife, Margaret – she gave me a cheque before they cut a single blade of grass – and his nephew, Jimmy Gray. Ken helped. Will and Peter Clark helped. Make hay while the sun shines. And for days it did. For the fields, it's the hair-dressing moment of the year. When first cut, the hay lies flat and shiny on the razored surface. The sun glints along it like a light on those snips of wet hair that lie on a barber's floor. To dry it, the grass is tossed with the tedder – Margaret's job, eight hours at a stretch, up and down, up and down in the battered old Ford 4000 tractor, 'stirring it about' and mussing it up, the shampoo shuffle. Then it is fluffed back into rows for the baler, the final hairspray and set. What this means is long, long hours at the wheel of a tractor, looking back over one's shoulder at the machine that's doing the job, with such concentration that Fred went past me three times before he noticed that I was standing there on the edge of the field waiting to talk to him.

Everything went like a dream and the hay lay soft, light and 'blue' as Fred called it, a green tinge to the grey of the drying grasses, in the rowed-up lines on the field. Not a drop of rain had touched it. This was some of the best hay anyone had made for years. But then, of course, things changed. The forecast predicted thunderstorms that evening and the baler broke. Fred had bought it from the two old Davis boys who were retiring from the place over the hill. I only heard this late in the day, but it was not surprising Fred had kept the source of the baler a little quiet. That was where, the year before, the BBC had stumbled on a fragment of old England, nettles growing through abandoned horse-drawn hay rakes, fields that looked as if they had just got out of bed, a farmhouse soft in its long slow journey towards dilapidation. They had decided to make their film of *Cold Comfort Farm* there. God knows how old the baler was. There was something seriously wrong with it, but no one could work out what. The bales it produced were either the size of a handbag or emerged 8 feet long, oozing out with a terrible constipated slowness from the machine's rear vent.

'Neither's any good,' Fred said, and for 15 hours, while the rain threatened, men from various parts of Sussex pored over its innards. The handbook was out on the field. They drank Ribena. Parts were greased, others rubbed down. 'If I never see another Case International baler,' Ken said, 'I won't be sorry.' The weather forecast was getting worse by the hour. In the end, there was nothing for it and the hay was baled in these stupid lengths. As soon as it was done, we stacked it on trailers and carted it into the barn, just as it was, the long and short of it, an acre an hour for six hours of exhausting, dusty, sweaty work. By the time the rain came, my 500 bales were in the barn, perhaps

£1,000 worth. We felt delighted. The hay was saved and the barn, filled to the eaves, smelled sweet and musty. We were all sneezing with the dust and seeds in our hair and nostrils. My forearms were pricked with the stub ends of the stalks. Motes danced in the sun where its light came in through the wide-open doors of the barn. The hay was stacked 15 feet high in two blocks, each 20 feet deep and 20 square, one block on either side of the barn's central cartway, held there by the diagonal oak braces and the central oak pillars of the barn. Scraps of hay lay on the barn floor, like scattered herbs in a medieval house.

The picture of fullness this gave me, a building as tightly stuffed as a pillow, a barn filled with exactly what it was meant to be filled with, was a version of the completeness I had come here for. The hay had been made – baler or no baler – with the techniques people had learned when they first kept animals in this country 5,000 years ago, and the five of us stood around drinking tea, looking at the completed thing as one does when a job is finished, all of us, in a way I can hardly describe now, jointly happy at what we had done.

I couldn't work out why Fred was looking so pleased with himself too. All his hay was still out in the fields, baled in the modern round jumbo bales, which a highly efficient brand-new machine had been creating all afternoon. They were bound to get a soaking. 'Oh, I don't worry about that,' Fred said through his one eye. 'Rain doesn't hurt jumbo bales. They can stay out there for weeks.' So why on earth had we sweated over our ridiculous salami/sliver-sized bales all afternoon? 'Oh,' Fred said, with a grin the size of the English Channel, 'I thought you wanted to do it up here like we did it in the old days. You didn't want those jumbo bales, did you? You wanted something you could get sweaty picking up and putting down so you could feel what it was like to be a real farmer. You did, didn't you?' I looked up at him as he asked me and saw – one of those moments of true recognition – that Fred had both of his eyes, the colour of the sky on a distant, sun-swept horizon, wide open, as the first drops of rain began to fall on the bleached and razored fields.

The hay was in, but the trees were suffering. For weeks on end, from midsummer onwards, they looked bruised and battered. A ride on the Northern Line in the evening rush hour would not have revealed a more exhausted set of faces than the trees displayed that summer. Our neighbour from Perryman's, the young dairy farmer, Stephen Wrenn, who had taken some of the grazing for his bullocks, came over for a drink one evening. 'I don't know a farm that's as lucky as this one with its trees. You'll look after your oaks, won't you?'

We had long talks together about what to do with this land. He had persuaded his father to give up the dairy herd, rent out the milk quota and turn Perryman's over to the new short-rotation willow coppice which can be harvested every couple of years and burned for energy. So the cows were sold and they were trying to sell his milk quota. But it had been such a dry year with so little thick growth in the grass – all top and no bottom, as they say here – that no one was in the market to take on extra capacity because feeding the cattle in the coming winter would cost a fortune. Drought was stalking all of us.

Even at the end of July, the leaves on the trees already looked used, dirty, in need of replacement. By early August, some of the hawthorn and hornbeams in the hedges were already largely yellow. By the end of the month, the spindle leaves were spotted black and had dried at the edges into a pair of narrow red curling lips. Elders had gone bald before their time and there were ash trees of which whole sections had been a dead manila brown for weeks.

An oak tree 60 feet high and wide may drink about 40,000 gallons of water a year. It is a huge and silent pump, a humidifier of the air, drawing mineral sustenance from these daily lakes of water that pass through it. Where, in a summer like this, could such a tree have got the income it needed?

The truth is, at least with some of the oaks here, they had been running on empty, trying to live through a grinding climatic recession. I was fencing between the Cottage and Target Fields – the Wrenns' bullocks had, as ever, been getting through – and I leant on a low oak branch as I unwound the wire. As I pushed against it, quite unconsciously, without any real effort, the branch, perhaps 15 or 20 feet long, came away from the trunk of the tree and dropped slowly to the ground. It had seemed fully alive, decked with leaves and new acorns as much as any other, but it pulled away as softly, as willingly as the wingbone of a well-cooked chicken. I pushed it into the fence, as an extra deterrent to the cattle.

Two days later, at the top of the Slip Field, I found an enormous branch, full of leaves and acorns, lying on the ground beneath its parent, perhaps 40 feet long, the bulk of a small house or a lorry. It too had been neatly severed at the base, as if the branch had been sacked, ruthlessly dropped for the greater good of the whole.

These living branches rejected in mid-season made me look at the oaks here in a new light – their scarred bodies, their withered limbs, the usual asymmetry to their outlines, the slightly uneven track taken by each branch as it moved out from the main stem – and started to see each oak not as a thing whole and neatly inevitable in itself, but as the record of its own history of survival and failure, retraction and extension, stress and abundance. Each oak has a visible past. The story it tells is more like the history of a human family than of an individual, forever negotiating hazards, accommodating loss, reshaping its existence.

One afternoon we were all in the kitchen together. We were sitting around the table and Ken as ever was regaling us with stories of past triumphs. Suddenly we heard, coming over the wood, from the lane that runs down from Brightling Needle towards the valley and on up to Burwash, the sound of sirens: ambulances, police, fire? We didn't know. It was a rare noise, more troubling here than in any city street. It marked a real person's crisis, that of someone you knew. We heard that evening. Stephen Wrenn had been killed. A tractor he had been driving toppled over a little bank, no higher than the back of a chair, and crushed his head. He died instantly. He and his new wife had only just returned from their honeymoon. The entire village went into shock over it. Two or three hundred people attended the funeral and the vicar who, a couple of weeks previously, had married him, helped bury him too.

One evening later that summer, when I was taking the children down to the seaside to play on the sands at the mouth of the River Rother, I happened to meet Brian Wrenn, sitting quietly by the river, looking out to sea. I sent the children on down the track and sat down next to him. We talked about Stephen. Brian said he was 'learning to face a different future'. It was as if his whole being was bruised. There is very little to say to a man who has lost his child.

At the edge of our land you could see, across the little side-valley of a stream that runs down to the river, one of the Wrenns' very banky fields. Before, it had been grazed tight, thistly and docky in patches like every bit of land around here, but with a background of new, bright green grass. Now, with the cows gone, and with Stephen gone, it looked different, the hay long and not cut until late, an air of abandonment to it, or at least of other matters on the mind. I looked across at that field and in it saw what had happened to the Wrenn family, the stupid, trivial, devastating disaster, the slice taken out of their lives.

I will always remember Stephen for the grinning optimism of what he said about the trees, the way we were lucky, blessed with the oaks here. 'Look after your trees,' he had said to me, and I will, as a memorial to him if nothing else. Isn't it a habit, in some part of the world, to plant a tree on a person's grave, to fertilize a cherry or an apple with the body? It seems like a good idea. That, anyway, is the picture I now have of Stephen Wrenn, but it is an oak, not a fruit tree, that is springing from his grave, the big-limbed, dark green, thick-boled, spreading, ancient kind of oak, so solid a part of the country here that it is known as the Sussex weed.

In the aftermath of Stephen's death, we were all rocked back. I took to spending time in the autumn wood. It is, on a quiet day anyway, a pool of calm. All the rush and hurry evaporates in a wood. If you lie down there, nothing happens. There is a sort of blankness, a consoling eventlessness about it. If a pendulum were swinging there, it would be floating as if on the moon, weightlessly falling, weightlessly climbing the far side. A wood distorts and thickens time. Occasionally, a small five- or seven-leaf frond off an ash tree, or a single hornbeam leaf, will spiral towards you. A pigeon, with a chaotic bang-shuffle to its feather noise, will fluster out of the trees.

Those are only the headlines; the body-text is absent. There is no busyness here. The extraordinary patience of these vegetable beings is what defines them. The way in which the trees stand and wait, open-armed, their leaves dangled in the air for sunlight, their roots spread hemispherically beneath them, capable of doing no more than accepting the wetness that might come to hand, this is a form of existence that could not be more alien to our own. The leaning patience of the tree, its long game: that's the beauty and the dignity of it.

As I lay in the Middle Shaw one morning that autumn, escaping work, fed up with it, haunted by Stephen's death, a sudden squall blew through the trees, unfelt at ground level but caught and noisy in the crowns of the oaks. It was a blast from the west and in that sudden wind, the wood began to knock and cannonade around me; the acorns, of which there were more that year than anyone could remember, were being blown and shaken out of their cups. The wood quite literally was noisy with the oak's seed rain, as the acorns bounced down through the lower branches and spattered on to the leafy floor. This was the seeding moment of the year, the culmination of the year's life. It was as near as a wood could ever come to breeding, the climax of the year. But all this rattling activity did not represent the reality of the trees. A true film of a tree's life would be grinding in its slowness, nothing but the great non-event of gradual enlargement. But that slowness is what is beautiful about a tree. Its concurrence with time, its superbly long rhythms, cannot be captured in a way that would make people watch or listen to it. The music of a wood would make Gorecki's Third Symphony look up-tempo, a sharp little dance-tune. Which makes the real thing so rich and so rooting if you manage to make the time to listen to it.

Or so it seemed that autumn. The wood was a balm-bath, a long slow statement, simply, of the trees' presence and persistence and dignity and life. That is the reason groves are sacred. Great trees stand as a reproach to our business, to our neurotic rush and hurry. But what a price they pay: incapable of defending themselves, as passive as whales under the harpoon. For all their dignity, they are no model for us. Anyone who acted like a tree would be thought mentally deficient. That is the conclusion I came to: we have to be anxious to be human. Passivity, calm and the long view: none of it's quite enough.

I knew I had to start getting a grip on the land we had now acquired. The woods were the place to start. Wooding is a winter job, when the sap is down and trees are not hurt by the saw. A winter-cut stool of ash, hornbeam, hazel or chestnut will sprout again in the spring, and those new fresh sprouts, called 'spring' in this part of the world, will re-establish the tree as a living plant. And so I asked Peter Clark if he and I together might begin to get the woods in order.

Peter Clark had been wooding for 14 years or so, all his adult life, and he was expert at it. He used his chainsaw like a balloon-whisk. A flick here, a zzzzzz there and order came out of chaos. There was a businesslike air to the way he approached the semi-derelict tangle of bramble and wind-blown tree. He didn't, as I would have in a half-hearted, uncertain and rather respectful way, nibble at the edges, trimming this, pulling away at that. He waded into the central problem. Confronted with the giant collapsed ash stools, the muddle of elder and bramble and old splintered oak limbs, he attacked them ruthlessly and systematic-ally. The cosmetics were left till later. Meanwhile, the stacks of usable cordwood grew at those points on the edge of the wood where, in a ground-hardening frost, a tractor and trailer would later reach them. His fires consumed the toppings, the useless bits and pieces. Every day that winter they burned in three or four places at once, positioned so that the smoke could chimney out through a gap between the big trees around them. From a field or two away the wood looked like a small leafy settlement, with the smoke climbing out from the three or four separate hearths and the chainsaw whining and relaxing, whining and relaxing as another fallen thorn or overgrown hazel was sliced and readied.

It was a wonderful sight – in the mind's eye as much as anything else – Peter moulding the wood in the way other people might pick up a lump of clay and shape a pot from it. He was a gentle and not especially gregarious or socially confident man. If there were other people about, he would often decide not to come in for a cup of tea or for lunch. Wooding is a private business, done in private, the results remaining virtually private, the whole event without a public face. And it was there, in that self-contained world, that he excelled. 'Do you like wooding?' I asked him and he replied in the way you might expect. 'It's a job,' he said and lifted his eyebrows into a smile.

We have four patches of woodland on the farm. One, the Way Shaw, is a field that was let go before the war and was now a thicket of bracken and wind-twisted birches. Ken said the remains of a V-1 doodlebug lay somewhere in there, but nobody knew where. Two of the others, Toyland Shaw and Middle Shaw, are old hornbeam coppices with some big oaks in them. The fourth, the Ashwood Shaw, is a wonderful old ash coppice, with giant stools growing on a steep bank between Great Flemings and Hollow Flemings, some of the stools twelve and fifteen feet across, with four or five 60-foot-high trees growing from each divided base.

This, in miniature, is a rich inheritance, an ash wood and a hornbeam wood providing the two necessary materials: one light but strong, making perfect poles for the handles of tools, for rakes and hay forks, the other tough and resistant. Mill cogs were always made of hornbeam wood and whenever I look at them I think of that, the iron hardness lurking under the oddly snake-like bark, the trunks not making good clean poles like the ash but twisted, fixed in a frozen and rather ugly writhing. The ash and the hornbeam, the calm and the perplexed, the classic and the romantic of an English woodland.

I was feeling my way with the wood. Clearing up was obviously the first stage of what to do here, but it wouldn't be enough. That autumn a couple of enormous ash trunks crashed out of the wood and into Hollow Flemings, the field below the shaw. There had been no great winds, nor anything else to disturb them. They had simply grown too big for their foundations. The leverage of the 60-foot trees became too much and they snapped out of their fixings at ground level, leaving a torn stump and exacerbating a weakness which meant that other stems from the same stool would soon go. The only way to save the plants was to cut them down. New growth would spring from the shorn stubs and the interrupted cycle of coppicing, which, judging from the size of the stools, must be many centuries old on that bank, undoubtedly a medieval landscape, would be resumed.

I talked to a local timber man, Zak Soudain, about the wood

and he was keen to have it. The bottom end of an ash trunk, where it moves slightly out from the stool and then up towards the light, a shape which preserves even in old age the first directions taken by the new stem in the first spring after coppicing, is the most valuable part of all. It is used to make lacrosse sticks. Nothing else will do. The rest, the straight clean lightness of the ash, goes into furniture.

So far, so profitable. But there was a hazard. We were overrun with deer. As we looked out of the bedroom window soon after seven in the morning, there would be eight or ten deer grazing in the field. The fawns in September were still playing with each other in a puppyish, skittish way. There was a stag with a single antler left, walking around lopsided like a car with one headlight out. Deer eat young trees. If we cut the ash down, they would chew off all the new shoots, the stools would die and I would have destroyed a small sliver of the late medieval landscape. But if we didn't cut the ash trees down they would probably collapse in the next big storm and the wood would be destroyed anyway. Deer-fencing was prohibitively expensive and ugly. I wasn't quite sure what to do about this and so I dithered while Peter easily and confidently moved through the fallen mess of things. I asked him one day what he would do. 'I don't know,' he said. 'It's not for me to say. You've got to decide, Adam. It's your wood.' I didn't tell him that, as far as I could see, the wood felt more like his.

That autumn I bought our first sheep: 20 Border Leicester ewes, which had already been through one year's lambing. They were advertised in the local free-sheet, £600 for the lot. I knew we had to plunge into livestock and this was a way of doing it. Will Clark and I drove over to look at them. Will said he knew about sheep and did quite a bit of squeezing of the back end of the animals in question. I certainly knew nothing. The woman selling them, wearing a fetching pair of buckskin chaps, said they were marvellous. So I bought them.

Carolyn Fieldwick, Will Clark's daughter and wife of Dave Fieldwick, the shepherd, had a ram to sell us. I bought him for £100. He was a big, stumbling, black-faced Suffolk and we called him Roger. He arrived on 5 November and started to mosey around our field full of ewes. If a ewe conceives on Guy Fawkes' Day, Ken Weekes told me, the lamb will be born on April Fool's Day. Roger seemed, it must be said, quite cheap at £100, and looked a little seedy. I could see him in a Dennis Potter play, snuffling around the ewes' rear ends like a tramp going through the dustbins at the back of a restaurant. They didn't much like the look of him or his intentions and used to move off to eat more grass in some other, less interfered with part of the field.

It brought back memories of 18-year-old parties, in which all the girls were pristine, self-sufficient and adult and I was a grubby, grasping bundle of unattraction, trotting around about 2 yards behind them. At least I didn't have to wear the sort of thing we put on Roger, a harness that Helmut Newton would have been proud of, holding a large yellow block of crayon wax in the middle of his chest. Whenever Roger managed to corner a ewe, he rubbed this, as a side-effect so to speak, all over her bottom so that we would know she'd been done. After the best part of a week, his score was two yellowed bottoms and one ewe that seemed to have an intensively crayoned left shoulder. Radical misfire or poor sense of geography: whichever it was, nothing could have been more familiar.

What an agony for poor Roger! So many requests, so much rejection. I caught him in successful action only once: a desperate five seconds of up-ended quiver and then down on all fours again, that look of hopelessness flooding back in, a sense of everything being over, a look on his poor, crumpled-ear face of utter bemusement. Why, I said to him, can't we all procreate like the trees?

Winter came sidling up on us. By mid-December, the darkness had lowered over the whole place, that terrible lightlessness when all you can do is remember the long lit summer, the after-hay evenings when the fields had a purified cleanness to them, patterned with an odd and unplanned-for regularity in the bales waiting to be collected, each of them throwing its shadow to the next, like a dabbed mark with a broad-bladed pen, while the dog is manically teasing some left-out wisps of hay and the children are playing man-hunt among the bales. What a sudden inrush of lost time that is.

My daughter Rosie, who was two that year, thought the trees were dead. 'The trees are dead,' she said one morning after break-fast, as one might announce that the war in Bosnia was over or Arsenal were third in the Premiership.

'Not dead,' I said, 'just resting.'

'Are they sleepy?'

'Yes, they are, I suppose.'

'Why aren't they lying down then?'

Anyone who doesn't believe in the reality of Seasonal Affective Disorder might learn a thing or two if they took a trip to the Sussex Weald in winter. Our own immediate surroundings that December represented the English winter in excelsis: a sapless, shrunken sump. I stayed inside as much as I could and averted my eyes from the windows as I passed. The mud lapping at the walls of the house on two sides had become a glutinated bog decorated with grey-eyed puddles and the semi-mangled remains of the rubbish which something was tearing open at night and distributing among the earth-heaps and trench systems. You could hardly blame the creature; no one could tell that scattering half-consumed, half-rotten rice-puddings and stock bones over what used to be the garden wasn't precisely what we had in mind.

The chickens we had foolishly acquired roamed delightedly among the old-food-encrusted earthwork-play zone where we let them out every day. They redistributed the mess. None of it ever seemed to disappear.

I had come to hate our chickens. They lurked about in the same murky province as unwritten thank-you letters and work that's late, the guilt zone you'd rather didn't exist. One is meant to love chickens, I know: their fluffy puffball existence, the warm rounded sound of their voices, a slow chortling, the aural equivalent of new-laid eggs, and of course the eggs themselves, gathered as the first of the morning sun breaks into the hen house and the dear loving mothers that have created them cluster around your feet for their morning scatter of corn.

Well, I hated them. Before the chickens arrived, I loved them. I sweated for days, building their run with six-foot-high netting, buried at the base so that the fox couldn't dig in to get them, with additional electric fencing just outside the main wire as another fox deterrent. I made a charming wooden, weather-boarded house for them, the inside of which I fitted out as though for a page in *Country Living*. There were some elegant nesting boxes, with balconies outside them so that the hens could walk without discomfort to their *accouchements*, ramps towards those balconies from the deeply straw-bedded ground, a row of roosting poles so that at night they could feel they were safe in the branches of the forest trees which the Ur-memories of their origins in the forests of south-east Asia required for peace of mind.

When it was finished, I sat down on the rich-smelling barley straw and smoked a cigarette, thinking that this was the sort of world I would like to inhabit.

We should have left it at that, but we didn't. We actually bought some chickens. And a cockerel. He came in a potato sack and when I tipped him out on to the grass and dandelions of the new run, he stood there, blinking a little, surrounded by his harem, and I couldn't believe we had acquired for £8 such a shockingly beautiful creature. He was a Maran, his white body feathers flecked black in bold, slight marks as if made with the brush of a Japanese painter. His eye was bright and his comb and long wattles the deep dark red of Venetian glass. He seemed huge, standing a good 2 feet high, and this fabulous, porcelain-figure colouring made a superb and alien presence in our brick and weatherboarded yard. His chickens, which he cornered and had with a ruthlessness and vigour we could only admire, were dumpy little brown English bundles next to him, heavy-laying Warrens, dish-mops to his Byron. For two days after his igno-minious sack-borne arrival, he remained quiet but then he began to crow, his cry disturbingly loud if you were near by but, like the bagpipes, beautiful when heard in the distance, down in the wood or with the sheep two fields away.

Within a couple of weeks it was going wrong. I was collecting eggs with my son Ben, who was seven. It was early evening and the chickens were still out. We didn't realize it but the cock was already in the house and with only the warning of a couple of pecks on my feet, which I didn't recognize for what they were, he suddenly attacked Ben, banging and flapping against his trouser legs in a terrifying explosion of feathers and movement and noise. Ben and I scrambled out of the hen house, him in tears, me shaken.

It worsened over the next few weeks. We were all attacked in turn until one Sunday morning found the entire family cowering behind the glass of the back door, checking to see if Terminator, or Killer Cock as he was also called, was out on the prowl. He had come, I am sure, to sense our fear and was now certain of his place as Cock of the Walk. He had to go. Of course, there was no way I could bring myself to capture him and so we hired a professional to take him away. We thought there might be the most horrifying execution scene in a corner of the yard. What actually happened was a lesson in the psychology of dominance. Alf Hoad is a man with enormously hairy arms. He lives in the village and shoots deer. He was our chosen executioner. Alf arrived in his Land Rover, stepped out of it carrying a sack, walked up to the cock and put him in it. My manliness rating dropped like a stone. The children now look on Alf as something of a god. He took the cock away alive and used him as a guard dog to protect his pheasant chicks against foxes.

It was a relief when Killer went. We could walk about again outside without fear of a rake up the back of the legs, but, without their man, our ugly little brown chickens suffered a drop in status. I looked at them and saw only the slum conditions in which they lived – my fault, they didn't have enough room – and their scrawny appearance – nature's fault, as they were going through the moult – and I blamed them for both. They stopped laying with the days shortening, and so we didn't even have any eggs. In fact, we were quite pleased about that because we had come to think eggs disgusting.

People, I now understood, had got the wrong idea about chickens: they are not the soft, burbly things they always appear to be in pictures and advertisements. They are utterly and profoundly manic. This whole short history had taught me an important lesson. There is something about the chicken which invites maltreatment. No one, I think, would ever have tolerated the idea of battery ducks, even if that were possible. People have caged billions of chickens in the most intolerable conditions because everything about them tells you that they have no soul. This is not to condone it, but it does perhaps explain it.

The chickens somehow made the winter worse, its awful unshaven stubbliness. The whole of Sussex looked as if it had been in bed with flu for a week. Its skin was ill and a sort of blackness had entered the picture, as if it had been over-inked. No modern descriptions of winter ever put this clodden, damp mulishness at the centre of things. People always talk about ice and frost and glitter and hardness and crispness and freshness and brightness and sparkle and brilliance and tingle. It's all nonsense. England is at sea and has sea-weather, a mediated dampness. That winter it entered our souls.

In a sea of unglittery mud and damp prospects, with things

unfinished, never unpacked or never started all round us, we huddled over our fires. Visiting friends were amazed at the mess. Our first year had come to an end. Was it, I still wondered silently, a mistake? Did we belong here? What were we doing here? Were we going to be happy here? Had we swapped one sort of unhappiness for another?

Why did we stay when so many others leave, just at this point? The euphoria, the bursting of energy from the bottle as it was first opened, had popped and fizzed and diminished and sunk, leaving only the still liquid in the glass. We were left with the plain fact. We had our work to do. I was writing for the *Sunday Telegraph*, columns about our life on the farm and others more generally about the politics of the early 1990s, the end of the Thatcher era, the John Major interval, the coming of New Labour, the political conferences, the disintegration of the Tory world, the expanding levels of hope that seemed to emanate from the Blair camp. I had a coffee cup emblazoned with the slogan, red on black, 'New Labour, New Hope'. It has been through the dishwasher so often that the words are illegible now. I was writing profiles of the political leaders, spending two or three days 'up close and personal', as it said in the paper with Blair, Major and the Liberal leader Paddy Ashdown, while writing a book, my first for several years, on the restoration of Windsor Castle after its fire in 1992 for which I interviewed hundreds of consultants, architects, builders, members of the Royal Household, curtain makers, gilders, wood carvers. It was a busy, engaged time. Life was starting to fruit again.

Sarah was making the garden, writing her book about it, looking after Rosie, doing her best to make the house more habitable. It was not a good building. The core of it was a two-up, two-down, tile-roofed and tile-hung cottage, of an age nobody could quite establish but perhaps built at the end of the 16th century. It had originally been entirely oak-framed but, like many such cottages, laid straight on to the damp clay. The lower limbs of the oak had rotted and been replaced with a mixture of iron-thick sandstone, probably dug from the ridge to the east where the hornbeam coppice now is, and brick from the works at Ashburnham a few miles to the south. Those two rooms downstairs were our dark original kitchen and the tiny sitting room in which a fire burned all day long in the winter and where we used to huddle in the evening, close to its flames. The smell of oak smoke filled the whole house as if we were curing bacon. Each room was about ten feet square and six high. I couldn't stand up in the old kitchen but the floor had been lowered in the sitting room and in there the top of my head just scraped the beams. From some old planking I made seats which fitted into the fireplace, one on each side. If you sat there, the knees of your trousers got singed.

Upstairs, in this oldest part of the house, was our bedroom – taller but with low windows looking out southwards over the garden, the Cottage Field and the wood – and a second bedroom where Rosie slept, which also had a little fireplace in it and a similar window. It could hardly have been simpler.

In the 19th century, this tiny square box of a house had had another bay added to it, with another room upstairs (now a scrawny bathroom) and a room downstairs which had a door out to the garden. Sticking out from this enlarged block was a converted cow byre, which was where the boys slept, and on the other side Mr Ventnor's 1980s addition, a non-functioning and frankly ugly extra sitting room with a huge and hideous fireplace. We never used it, preferring instead that sheltering core, the old heart which felt like a good and safe place.

It was not entirely right as a house: no very good rooms, unusable in parts, over-large in others. We tried our best to make it better, moving the kitchen from its original site to the 1980s addition. Jake Farman, Lou's husband and an old friend, made a big fitted sideboard with giant drawers and over-thick shelves, like a child's idea of a dresser, stretching the length of the new kitchen. A cheap version of an Aga, called a Nobel, was installed at one end. Ken Weekes built a copy of a fireplace I had seen in an Italian kitchen at the other end, with a big plastered brick hood over a flagstone hearth raised about 3 feet off the floor so that you could sit on it, next to the fire, and cook over the glowing embers.

This looked good but it smoked and so we had to install an extractor fan on the top of the chimney, which whirred away like a toy helicopter trying to take off. Within a day or two of putting this up there, an old basket we were burning on the fire wafted itself up into the chimney and set the whole shaft alight. The firemen from Burwash – it was a Sunday lunchtime – put it out with a wet towel on the fire, the steam of which went up the chimney and dampened the flames. Nobody died and we all drank beer afterwards while the sun poured in through the windows.

The interiors of Perch Hill were frankly neglected. The loo upstairs had no lid. We hadn't replaced the old, grey and grubby carpets. Mr Ventnor's 1976 gilt downlighters and pine-louvred cupboards remained where we had found them. From the beginning Sarah and I concentrated on the outside and left the house till last. Why was that? Perhaps because the inside of a house is not quite of the essence. It is too changeable, too often changed, to have anything to do with the nature of a place. Interior decoration is pure imposition, while the reason Sarah and I came to live here was not to impose us on here, but here on us. And getting the outside in shape seemed like an attempt to bring out the virtues of somewhere like this: its vitality, depth and rootedness, its optimism and life. Wallpaper never flowers or fruits, does it?

Only in my workroom did I make it very different. I had lumps of the natural world littering my desk. Along with the computer discs, the Post-its and the unanswered mail lay a hunk of ironstone from the ground in one of our gateways, the iron clotted in the ginger-coloured stone like fat in pâté; a slice of limestone picked up on a beach in Ithaca; a pitted volcanic pebble from the Azores; and another from the same islands, a drop of lava thrown out by a volcano that erupted in September 1957, the month I was born. It sits here now in front of me like a blackened bun, smooth on one side as if with a crust, rough on the other like torn-off dough. There's a flanged iron gate-pintle, made from Wealden iron and found burnt in a bonfire so that its surface is blotched an oxidized red, as though its skin were diseased; a flint from Dungeness, near Derek Jarman's garden; a basalt cobble from an island in the Hebrides; and a clay pot, made out of clay dug from the orchard here, fired in the oven and then washed by mistake in the washing-up machine so that it looks as if it has been recently excavated from an Iron Age hut.

I use most of these things as paperweights, when the wind blows in from the open door, but they are also more than that. They are, quite literally, touchstones, elemental souvenirs, carried back from trips abroad or walks in the fields. They are good for many things but above all for their lack of domesticity, their out-of-placeness in the tamed world inside. Their surfaces are cruder or more basic than their surroundings. They flake and chip. They are, on the whole, much heavier or at least denser than most of the things you find in a house and both their holdable weight and their retained cold on a hot day seem to stand as evidence of their worth. They belong to a serious world outside the light conveniences of a domestic existence. The keyboard and the PC, the telephone and answering machine, the printer, the modem, the pens, the drifts of paper: none of these could outlast the talismans alongside them. They are the ballast.

The curious aspect of these gathered but still naked objects is the way they are more significant here than where you found them. When you put them in a room, they summon, without effort, wide acreages of other worlds and previous times. I can look at the battered clay pot and bring to mind the summer night I dug its clay, with a friend, a potter, from the orchard pit, and our intermittent, half-embarrassed conversation. The volcanic Azorean bun still has about it the air of warm, soaking mid-Atlantic rain the morning I picked it up on the volcano's ashy slopes, while the blackback gulls hawked and wheeled around me and the swell ate away at the shore. The Dungeness flint still comes from an early summer morning, with the light reflecting off the Channel, so that the whole sky was lit up as if in a studio. The sea-cabbage was in flower and the valerian and viper's bugloss swung in coloured drifts across the successive ridges of the shingle. Unlike everything else, from which memory and detail fades, it is almost as if the longer you hold on to these things, the more the associations and the sharper the recollections that gather about them. They are tangible memories, objects to deny time simply because they collect and store memories like filings around a magnet.

Why do we like these fragments of the landscape in our rooms? Perhaps having them is an intuitive recognition of our distance from the natural, a pitiable piece of magpieism, collecting flakes of the unadorned with which to re-balance the electronic, the centrally heated and the chronically tense state of modern life.

In the other rooms bits and pieces of furniture from Sissinghurst and from Sarah's mother's house stood around. We bought some rugs. An old bit of tapestry that had belonged to my grandmother hung in the little sitting room beside a French faux tortoiseshell cabinet that must once have been at Knole, the great Sackville house in Kent where my grandmother had been brought up. It was a ragbag but I only had to go outside, see the beginnings of Sarah's garden, help her paint the stripy gondola poles with which she was decorating it or lie down in the buttercup and sorrel salad of our incomparable summer fields, to understand why we were here, why this world of mess and inadequate money was the choice we had made. Mess and sweetness; the immediate life; tenderness; the joint enterprise; Rosie in our wake; those fields; the deer out of the window in the early morning; the dog playing with the sleeves of my jersey on the kitchen floor: these were enough.

That is why we stayed, because this place for all its failings represented what we had between us. Perch Hill was what *we* were. It was jointness, the embodiment of our first person plural. If we had left it, we would, it seemed then, have been leaving ourselves behind. Anyway, nothing was so visibly unfinished, raw and unachieved, a project still to be addressed. To have left then, before we had even engaged with the realities of the struggle to make it good, would have brought on a shrivelling of the spirit from which neither of us, I think, would have ever recovered. Mess, if you look at it right, is stimulus, chaos an invitation to make good, jointness the lifeblood we needed.

Patrolling the Boundaries

I BECAME obsessed by Kipling that winter. Bateman's, his house, a mile from ours across the fields, now belongs to the National Trust and I asked them if I could write the text to a new guide-book they wanted there. They agreed. I didn't tell them how much I had come to see Kipling's own position as parallel to mine: he was almost exactly my age, thirty-six, when he came here, in retreat from the world, and we were both in grief over lost children – his six-year-old daughter Josephine dead from pneumonia, my sons severed from me, or part-severed anyway, by divorce – and looking *for the sustenance an ancient landscape can provide.*

The boys came here often, regularly, and I longed for them to think of this as home. But I was under no illusion: some gap, like the cracks that open in drying clay, had appeared between us. My home was no longer their home. Sitting for those weeks in Kipling's study, at his desk, investigating and exploring his books, handling his objects, walking his land that all but bordered mine, reading into it the moral and emotional structure of his stories, I was wrapping myself in this Sussex Kipling, not the drum-banging imperialist but the haunted man in whom so much of my own predicament seemed to be prefigured.

Morning after morning, I walked over the hill to Bateman's. It is down in the valley, away from the road, away from the village, surrounded by hedges and high walls. Even though the Kiplings knew it to be a gloomy house, already with the air of sadness it still has in its rooms even on a sunny day, it looked in 1902 like the haven they needed, almost a fortress, stony-faced, protective, and they jumped at it. It was a house in which they could pull up a drawbridge behind them.

There was more to it than simple anti-sociability. A passage in *Something of Myself*, Kipling's late autobiography, provides the key. He is describing his own magical practices as a lonely boy, suffering in a boarding house in Southsea, separated from his parents and his beloved *ayah*, still in India:

> When my father sent me a *Robinson Crusoe* . . . I set up in business alone as a trader with savages . . . in a mildewy basement room . . . My apparatus was a coconut shell strung on a red cord, a tin trunk, and a piece of packing case which kept off any other world . . . If the bit of board fell, I had to begin the magic all over again. I have learned since from children who play much alone that this rule of 'beginning again in a pretended game' is not uncommon. The magic, you see, lies in a ring or fence that you take refuge in.

That last sentence could be a description of Bateman's itself, of the place of this valley in Kipling's imagination and of what it meant to me too. It is the magical zone into which others do not intrude and whose power and secret relies on a vigilant patrolling of the boundaries, on a perceived isolation in which the richness of privacy can flower. His American wife, Carrie ('a hard, capable little person', Henry James called her), was the Keeper of the Gate and there is a story remembered in Burwash, and told to me by Graham Jarvis, the butcher there, from the Kiplings' later years which dramatizes particularly sharply that exclusion of the world.

One of the Bateman's calves was to be slaughtered not for its meat but for its thymus gland. This gland is at its largest in young animals and is still thought by some to contain life-enhancing, vitality-inducing juices. After John Kipling went missing at the Battle of Loos in 1915, his father suffered repeated and acute gastric pain. The calf's thymus may have been intended to alleviate this. After the gland had been extracted from the animal, Carrie insisted that the rest of the body should not be used for meat. It was to be buried not on the farm but inside the garden, and the ground over it raked to a fine tilth which would show any disturbance. After that tilth had been prepared, she signed the raked ground with her own name.

This is an odd and disturbing image: the life-giving calf, dead and signed for in the garden, locked away from the rest of the world by a mother who had seen the death of two of her three children and shown no public pain. It seems like an unconsciously magical and demonic act, a sacrifice in a Sussex garden, a thousand miles from the world of Burwash and the straightforward use of animals for meat but rather near to the world of Kipling's own poetry and its sense of the enigmatic undercurrents flowing everywhere beneath the surface of the ordinary.

The more I read, the more it became clear that Carrie's protectiveness was also obsessive. In the later years she would not let a single piece of Kipling's handwriting leave the house; everything he wrote had to be typed out by the secretary. When she found the manuscript of *The Irish Guards in the Great War*, Kipling's long, careful testament to the men in whose company his son was killed, about to be sent to the publisher with some last-minute handwritten alterations by the author, she insisted that the entire text be retyped. Kipling himself had to apologize to the secretary.

But Carrie's ferocity was an act of love. It allowed Kipling, among many other things, to explore, in private and in all its ramifications, the place where they lived. The landscape and those who were bound to it became the heroes of the stories he wrote at Bateman's, at the long walnut desk where I sat reading them. Stone mullioned windows look out on two sides: one eastwards along the pastures of the valley where this author-landowner would have seen his two herds of conker-red Sussex beef cattle and the Guernsey dairy cows; the other, to the right, over the woods that clothe the sides of the valley, the woods from which Puck emerges in the stories like an earthman-impresario to conjure magic for the children, and into which, at the end of each story, he melts wordlessly away.

The figure of Hobden, the hedger and ditcher, the archetypal Sussex man, whose generations have been there for ever, and who knows everything there is to know about the place, is Puck's human equivalent. He is Sussex made flesh, the ancestor of all the Clarks, Groombridges and Weekeses who were also peopling my world. Kipling, as he wrote in a poem, may claim to be the proprietor of a wide stretch of the Dudwell valley, he may hold the deeds, but Hobden, even Hobden the poacher, *possesses* it:

I have rights of chase and warren, as my dignity requires.
I can fish – but Hobden tickles. I can shoot – but Hobden
  wires . . .
Shall I summons him to judgment? I would sooner
  summons Pan.

('The Land', 1917)

Kipling, the valley and I clustered together. We became each other's. He put his mark on it, in field after field, at the corners of woods and the twisting of the river, at farms and at cottages now already in ruins and mossed over. In return, the valley shaped his imagination. I trudged after them both. More and more that winter I walked in Kipling's world. The words he used to describe his relationship to it were 'wonder and desire', twin attitudes, one distant and admiring, the other distant and longing. Those were mine too. Working it out with map and text in hand, I found where Hobden's cottage and his forge had been. An alder carr now grows over what must always have been their sludgy, boggy site. At 'the sadder darker end' of the wood, further along the valley, I found what he had described in *Puck of Pook's Hill*: 'an old marlpit full of black water, where weepy, hairy moss hangs round the stumps of the willows and alders. But the birds come to perch on the dead branches, and Hobden says that the bitter willow-water is a sort of medicine for sick animals.' I, too, found myself there, suddenly surprised one day by this: ' "Hst!" he whispered. He stood still, for not twenty paces away a magnificent dog-fox sat on his haunches and looked at the children as though he were an old friend of theirs.'

For the whole Kipling performance here, Puck, the little brown pointy-eared earth god, is the master of ceremonies. He acts as the compère, smoothly and coolly emerging from the leafy wings, presenting the children with the astonishingly immediate and real past, and then just as deftly slipping back into invisibility. In his hands, the boundaries between the real and the imagined are dissolved; the strange and the alien are slickly wafted into concrete existence and with equal panache swept away. Each story ends with that quiver of closure and each new one begins with a sudden unapologized appearance of the strange.

There were multiple layers in this valley for Kipling and he provided another for me. It became for him, as it had for me, a sort of reservoir of the English spirit which can emerge from the leafy shadows for an hour or two and then slide back into them, a place where the gates are down between the landscape, the idea of history and the sense of other lives and other spirits inhabiting the world we call ours. I have walked the abandoned roads with Kipling's most famous lyric in my mind:

> There was once a road through the woods
> Before they planted the trees.

It is underneath the coppice and heath,
   And the thin anemones.
   Only the keeper sees
That, where the ring-dove broods,
   And the badgers roll at ease,
There was once a road through the woods.

Yet, if you enter the woods
   Of a summer evening late,
When the night air cools on the trout-ringed pools
   Where the otter whistles his mate . . .
You will hear the beat of a horse's feet
   And the swish of a skirt in the dew . . .
   As though they perfectly knew
The old lost road through the woods . . .
   But there is no road through the woods.

                ('The Way through the Woods', 1910)

That is still there too: the numinous haze above the leaf-litter on the wood floor; the moss-walled trenches of the old lanes dropping to the river where the shallow gravel turns it into fords; the black pools over which the willows and alders curve their long, flexed limbs like the struts of a tented dome; the knowledge of something having just passed, its scent hanging there in the way the smell of fox stays on in a sheltered hollow; and that fearful sensation when you find yourself flicking your head around behind you, knowing you are alone but sensing something else, a crack of a twig, a movement in the trees unexplained by the wind, the moaning creak of one trunk against another. All that is there, in fragments, and never more than at those marginal times, the early, dewy winter mornings and the ever-earlier evenings, as the sun comes in low and pale, its colour diluted by the damp in the air, washing the inside of the wood with sunlight it hasn't seen all year.

I relished all that privacy, the protectiveness, secrecy and subtle explorations of 'the things that are beyond the frontier', but I wondered too if the shut-awayness, the closure, the brusqueness of the Kiplings' dealing with the rest of the local world, whether that was really enough. The story of the buried calf, lurking in my mind for the weeks after I heard it, changed the way I saw that part of the valley. It coloured the landscape like a stain. It somehow drained it of blood. It denied one of the best and richest things about living here: the neighbourliness of it, the net of people we had already become connected to, a net which existed like a map overlying the physical map. Ken and Brenda Weekes, who lived in the neighbouring cottage and who told us how things had been before we even knew Perch Hill existed; Fred and Margaret Groombridge, who knew in their bones what this farm needed and just how we should set out on the task; Will and Peter Clark, who arrived in our lives virtually attached to Perch Hill; Nipper and Kitty Keeley, who both bred the dogs we loved and ran the great oak timberyard just south of Dallington on the way to Brown Bread Street, from where we got our wood; the Wrenns, who had suffered such a grievous loss in the death of their son Stephen; and the Fieldwicks, who were steering us towards looking after sheep more properly.

These were people we had come to know and rely on now. You could tell from their names that they were all what they called themselves, 'Old Sussex'. On the farm, each of them would tell me what to do and how to do it, how not to and what I had done wrong so far. There was a form of decorous, reticent generosity about the advice these men and women gave. They didn't want to tread on toes, but what they said always came with a slight and joshing twist on the end of it.

In one of these conversations with Ken Weekes that winter – it was about whether to put nitrogen on the grass – Ken turned to me and said, 'You know, Adam, it's very nice that you've come to live here.'

'Oh,' I said. 'Why's that, Ken?'

'Because it looks like you've got money to burn.'

For all that, Ken was a stalwart, helping us out time and time and time again. Neighbourliness is central to the way he is. Remembering his deep and social generosity of spirit, you turned to the cold picture of the dead calf buried in the garden, the anti-neighbourliness, the grief in it, and Bateman's no longer seemed to belong to the country in which it was set. It had become a disconnected anomaly. 'England is a wonderful land,' Kipling wrote to a friend just after arriving at Bateman's. 'It is the most marvellous of all foreign countries I have ever been in.' That remained true of his place in Sussex: never at home, in the way the Weekeses are, but consciously and repeatedly searching for a home, for a feeling of embeddedness in the place. The very search set him apart. Perhaps he was no more than an early forerunner of all those of us who have left cities to find a rural place.

As a boy at Sissinghurst I had known a wonderful rural province to explore and roam through, but it had been disengaged. I had never come close to animals or work on the land. Since then, in the world of books and newspapers, I had lived an urban or at least an urbanized life. So for me this was more than a return to the source: it was an attempt to dive into the source at a deeper level than I had ever known.

But there is a problem with this pastoral drive. Pastoral – the idea that a rural existence can somehow regenerate those who give themselves over to it – carries the seeds of its own failure. It is, by definition, a sophisticated attitude. Only those who have abandoned the idyll, or have had the idyll withdrawn from them, are in search of it. The fact that they are searching means that they cannot find what they are looking for.

Even in the knowledge of its inadequacy, though, pastoral seems to me not a worthless but a necessary myth. It provides a sanctuary in which a bruised mind can rest. It puts a torque on the material concerns of the everyday, twisting them towards something else, some better state.

With Ken Weekes one day, as the whole Kipling phase was nearing its end, the neighbourliness between us took a step forward. He had a gun for sale and I bought it, on an impulse. It was a 12-bore Silver Sabel De Luxe Side Lock Ejector made by a Spanish gunsmith called Gorosabel. 'That's £1,800 new,' Ken said. 'You can have it for £540, seeing as you don't know what you're doing.' I'm not sure why it was £540 but I'm putty in Ken's hands and so I said yes and for the first time in about twenty years I had a gun of my own.

Guns smell delicious: mineral, acrid, serious, not pretty, not part of any cocktail-party scene. The smell of a gun is uncompromised. There isn't a soft edge to it. Someone should think of marketing gunsmell as an aftershave.

My Sabel De Luxe was a beautiful thing for the few days that I had it – chased on the silver plate that covers the side lock mechanism, the stock polished, dense, rich, reliable, the opening mechanism sharp, the pleasure of cameras in the precision of its click open and its snap shut. It combined the best of furniture and jewellery, intricacy and solidity, designed for a purpose, moulded to the body, made for me. Of course a gun is a glamorous thing.

I went out with it early one morning, late that February. It was strangely warm and the grass was growing as though it were May. Custard's snout was dripping from the dew bath, like a boxer with his sweat up. Every time he shook his face, the wet flew off it in a halo.

Noises in the distance are more audible first thing in the morning. There was the traffic on the Burwash–Heathfield road along the ridge; jets into Gatwick; pheasants over in Leggett's Wood, a cousin noise to our own stupid chickens clucking; the hollow knocking of a sledgehammer on a chestnut fence post down near Bateman's; the Coxes' tractor at Sheepshaw Farm, grumbling and farting into life; the spatter of something falling in High Wood on to the leaf floor below the trees.

There I was, standing in the middle of this near-silence, with the sunshine breaking in bars on to the grass of the Slip Field, the warming, smooth-skinned gun in my hand and the dog, obedient, wet and bored at my feet. I stood in the shadow of an ash tree and waited for the rabbits to emerge. I had seen them here before and throughout the grass of this little wood-lined sliver called Hollow Flemings are the nibbled patches and the scattered droppings they had left behind. If there was going to be a Perch Hill killing field, this was the one.

When I was a boy I had a series of guns at home in Sissinghurst: a BSA air rifle, with which I shot out the windows of the granary next to the barn and killed a robin at three yards on Christmas Day; a .410 shotgun; and then a Holland and Holland 16-bore, a beautiful, slim and elegant thing, more like a walking cane than a shotgun, with which I felt completely at home. I stalked about the fields of my stepfather's farm in Hampshire with my pockets full of cartridges and my heart full of bloodlust, popping off at this and that for the hell of it, enjoying the jolt into the shoulder and the *ding* in one's ears afterwards. It might be a rabbit one day, some pheasants or pigeons or squirrels the next. I wasn't much interested in the bodies or the eating of them. It was the shooting that counted.

This was the time we were all having to read *Les Caves du Vatican* for French O-level, with Gide's thrilling idea of the *acte gratuit* – one never translated the phrase – an entirely gratuitous action, for some reason always violent and nasty, to show that you were one of the cool gang, *les subtils,* not subject to the dreary rigidities of conventional moral life under which the others, *les crustacés,* suffocated. In the book the *subtils* shoved strangers out of train doors at 80 miles an hour; I murdered animals. My mother thought I was learning country lore and the Ways of Nature;

I knew I was living out the fantasies of a gay French intellectual half a century earlier. Life in the Home Counties had never been so authentic. If I'd been born to different parents in a different class, I would have been sent to Borstal.

The last thing I shot on my teenage grumblings around the Hampshire fields was a hare. It sprang up 15 yards away, out of the footings of a hedge, one December morning. I swung round on to it and shot it once. The shot had no effect and the hare went tearing on across the wide spaces of the field. I watched it for a while and then turned away, to put a new cartridge in the gun. Just as I did that, I saw or thought I saw the hare stutter and tumble over, as though a wire had tripped it. I walked over to the far side of the field and there it was lying on the grass. I touched its eye. The hare was dead. When we came to eat it, there wasn't a single pellet in its body. I'd missed it, but the noise of the gun going off had been enough. The hare had died of shock. I don't remember taking the gun out of my stepfather's cupboard again.

All that was 20 years before. In the meantime, I had shot a single stag, out of curiosity and after a long and beautiful stalk through the hills near Rannoch Moor, but that ended in absurdity too. It was a warm day and the herd of deer was sitting calmly on the cheek of hillside below the stalker and me, 100 yards away. Whispering, he told me that the stag I was to shoot was the one in the middle, 'the hummel'. A hummel is a eunuch, with no antlers. It had never been like this in John Buchan. Not only was my stag a castrato: it was fast asleep, dozing in the autumn sunshine. The stalker began by whistling softly to make it stand up. No good. A light clap of the hands – nothing. Finally, after some banging and shouting, the other deer got up and walked away but my chosen hummel, my sweet fat little eunuch, continued his afternoon dreaming. 'Ach, there's nothing for it,' the stalker said. 'Shoot it where it is.' So I did, through the neck, severing the vertebral column. The hummel never woke for his death; his head slumped forward on to his chest and from a distance there was no telling he had died.

These were the scenes running through my mind that morning as I stood in the fringes of the wood with Ken's gun in my hand. The sun climbed higher, the rabbits came out into the field but I didn't raise the gun to meet them. 'What is the point?' I said to the dog and we walked back up to Ken's to return the gun. 'I know what you mean,' he said. 'I only shoot clay pigeons myself nowadays. I don't know why, but I don't like the other any more.'

Another killing question loomed: the hunt. Kipling had banned it from his land and we did too. It was casually done. They asked if they could and we said they couldn't. I wonder now, years later, if this was too thoughtless on our part. I didn't think carefully enough about it, the casual disruption to other people's lives which our casual refusal involved. Dislike of the hunt was largely a gut reaction, founded, absurd as it might sound, on a dislike of horses. There is something in particular about a horse's bottom which makes it difficult to take the horse world seriously. Horses are not alone in this. Different styles of bottom will always colour the way in which you see different parts of the animal kingdom: the encrusted chaos at the back end of a sheep, usually the farmer's fault, too rich a diet, but essentially sheepy nonetheless; the pert neatness with which the rear of the cat meets the outside world; and the way dogs divert one's attention from the bottom itself with the wonderful flag-waving gesture of the tail, leading one's eye up and away to the tip-of-the-wag point of delight. That is the difference between a cat and a dog: one is a take-it-or-leave-it, I'm-no-one-to-make-any-form-of-social-compromise sort of animal, the other nothing but love and diversionary tactics.

Horses, though, are the apotheosis of the bottom. I cannot help thinking of them as mainly bottoms with a head stuck on the front. And what heroically enormous bottoms they are! That pair of huge round buttocks, a couple of Brobdingnag conkers, sheened by their owners like a schoolboy with his cherished, oven-hardened seventy-sevener, so full, so fleshy, so bottomy, even when at rest: that is the essence of the horse.

Prizes are awarded on racecourses to the stable lads who devise the most elaborate chequer pattern in the coat of the horses' bottoms – called 'quarters' in this context – and that, I think, is some recognition of the importance of the bottom to the whole being of the horse. But it is when they set off on their first trot of the morning that the horse bottom comes into its own. This is the rhythmic, stomach-easing, first stretching out after the gases have been accumulating all night and it is then that the massive bums, the most buttocky buttocks in the animal world, like a chorus of podgy ballerinas performing to a loving audience in front of them, breathe a series of little synchronized farty kisses into the morning air. Squeeze, squeeze, squeeze, fart, fart, fart: it's the flagship gesture of the horse's existence. And the farts, of course, are only the preliminary to another synchronized movement, the flopping out of those steaming piles of Loden-green dung, flop flop, flop, one after another. How the horse manages to make them look as if they have been created by some kind of internal ice-cream scoop, I have no idea. It is one of the miracles of nature.

Riders, I think, know this instinctively about the horse and the bottom. It is why, in their manner, they attempt somehow to compensate for the inherent absurdity of their position. The sillier the bum, the more serious the person on top of it. The pretence doesn't work. To anyone not on a horse, anyone on a horse looks stupid. And the more they try to look dignified, modelling themselves on the equestrian statue of Marcus Aurelius that used to stand in the Campidoglio or whatever it might be, the more you see some idiot looking like the cherry on a Bakewell tart. And if the horse then goes into its early morning intestinal exercises, the effect is worse.

This, I think, may have been at the heart of the difficulties that hunting was going through in those years leading up to the ban in 2004. It is nearly impossible to take someone who is up on a horse seriously, and if you can't take them seriously you can't really accept the idea that they should be horrible to foxes, even when foxes have been as horrible to your ducks and chickens as they had to ours.

That January, the fox had already nabbed a couple of ducks and a day or two later had got at the chickens. In the morning we had eleven, poking around the mist-sodden winter fields and looking happier than they had for a long time. Freedom suited them, to a degree anyway. By the afternoon, I could only see seven. Quite often in the cold, they used to go into the barn and shuffle about in the mad Burmese jungle-fowl style their genes instruct them to adopt, flicking at the hay and pretending they are deep in the tropical rainforest. I did find two in there but two were still missing.

Up in Jim's Field, our best and biggest hayfield, I found the other two dead in the grass. Dave Fieldwick's sheep, with his hideous, expensive, Tyson-lookalike Texel rams, were ignoring the two dead animals in their midst. I don't know if a fox had done it. Perhaps a dog had taken and dropped the chickens here. The grass had been pulled at and tufted by the sheep so that it had the look of hair first thing in the morning, a chaotically mussed pelt. Only spring would restore a sleekness to it.

On the grass, whose colour was now halfway between green and tawny, the two dead chickens lay, 10 yards apart. One was whole. I couldn't see a mark on it. The other had a big deep gash down its chest, a cut that laid back the flesh in the way that a butcher's knife scoops the raw chicken breast back from the raw chicken bone. Chest or breast? The sight in front of me hovered between the two. It was shocking somehow, and I was surprised to be shocked, that the actual cut body of the dead chicken looked like 'chicken', the stuff you see in the supermarket chill cabinets, sitting in its polythene tent on its expanded polystyrene tray, its leaking juices being mopped up by its pinkish poly-something nappy. Of course I knew 'chicken' came from chicken, but I had never had the fact pushed in my face before. I eat 'chicken' all the time, but the idea of eating these recently murdered things seemed, in a way I can't quite fathom, impossible.

I picked the bodies up by their smooth and scaly legs, that part of a bird that makes its reptile relatives seem so near to hand, and flung them into the edges of the wood, up over the fringing hawthorns and into the oak and hazel scrub beyond. Let the foxes feed, I said to the corpses arcing outward through the air, and turned for home. The bodies thumped to earth behind me. Their sisters were fluffing up their feathers into powder-puffs against the cold, those little legs sticking out below them like the wire stands on which shoes are sometimes displayed in high-class Bond Street shops. For the first time, I felt sorry for the chickens, victim-creatures, the huddled mothers of children they would never see, not really at home here, their genes pining for the jungle where they belong. Perhaps, I thought, I should take our chickens to some stretch of Javanese wilderness and release them there.

Despite that experience, I had no animus against the fox – of course I preferred foxes to chickens – and no love for the hunt. A week or two later, we had our first slight problem with them. They met just a mile away, next to the nineteeth-century obelisk known as Brightling Needle. It was a cold morning with mist in the valleys and everything was as it should be. Bottoms emerged from boxes, superbly bottomly, farting as they descended the ramps. Men in tweed jackets smoked Benson and Hedges at 7.30 in the morning – a hard drag followed by the classic Terry Thomas teeth-clench – and square-jawed women appeared in bowler hats. These ladies' heads were perfectly symmetrical about the bowler-hat brim, a living version of those drawings by Rex

Whistler which look exactly the same if you turn them upside down.

The traditional two policemen were there in two police cars. Parked on the side of the road was the traditional ageing, navy-blue Austin Maestro, slightly rusting on the sills, containing the traditional three antis, one with the traditional part-orange, part-bleached hair, one with the traditional row of earrings up the rim of his ear and one with a slightly innovative Barbour and tweed cap. I crouched down next to the fugged-up driver's window. It was a sign of the maturity of social life in East Sussex that the man with earrings said that the antis had 'a very good relationship with the hunt' and that things had gone 'very well so far this season'. It sounded positively parliamentary.

One of the policemen took down the number of the car and the occupants sat inside, smiling at him. The riders, some of whom I knew anyway to be nice, generous, open, subtle and supple people, looked stiff and stupid on their horses. Has anyone in the saddle ever managed to address anyone on the ground without exuding the whole 'my good man' aura that irritates and alienates so much?

I went home when the field plop-plopped off down the lane, with the hounds effortlessly elegant and dignified in front of them. Mid-morning we had the problem. There was a man on a horse standing next to the farm gate. I stood there looking up at him on his shiny bum and I felt like Wat Tyler at the gates of London with the men of Kent at my back. 'I am *so* sorry,' he said, that 'so' quivering with three more syllables in it than usual. 'We've had a great time and some of us seem to have got *so* overexcited that they just didn't see the electric fencing you've got up for the sheep and I'm afraid they smashed *straight* through it. I'm *so* sorry. We've put it all back together and it's fine now, and the sheep of course they're fine, but God how stupid can you be?'

Poor man. He was trying as hard as he could, but there was a structural problem here. It was toff up/peasant down, a spatial metaphor of everything you most resent. It was a medieval moment. Endless hours spent reading about bastard feudalism and town charters could teach you less about the Middle Ages than this simple confrontation. The horse creates an appetite for democracy and whatever the facts, the *feel* of hunting on horses is not democratic. Few people have felt the same sort of loathing for fishing or shooting – those crucially horseless versions of rural killing. The horse is to blame for the condition in which hunting finds itself. Everyone hates being talked down to.

One day soon afterwards I was summoned to London for lunch with a man who wanted to convince me that hunting was not what I imagined it to be at all. Rules, the delicious restaurant in Covent Garden which is usually full of fat red men in suits drinking claret, makes something of a cult out of eating wild animals. A little brochure sits on each of the tables to tell you why this is a good idea. Its basic sermon is this: wild animals are good for you because they spend most of their lives in the gym, migrating here and there, running this way and that, lean, not fatty like your lazy old farm slobs. Wild animals are more admirable than that, always on the go, never taking time for a proper lunch, lean achievers, career creatures who make Elle Macpherson look like a pig that has junked out on pork scratchings.

'Wild salmon will have swum the Atlantic,' it says exhaustingly, 'and so will have firm muscles, less fat and a varied natural diet.' No spare tyre on a wild salmon. All nature is a workout, with the best possible organic niblets as the reward. Wild duck are 'truly free-range birds', sea trout eat only the finest pink shrimps, grouse taste of heather and snipe of bog or, as this brochure put it, 'sweetly rotting wild mushrooms'.

These animals are what they eat, you are what you eat too and so if you eat them, in a sort of apostolic succession, you will become an elk. Magic. There is no need to think of anything as disruptive as actually taking any exercise yourself. You can stay in

Rules, you can go red in the face, you can tuck into a capercaillie on fried bread, larded with strips of woodcock freshly braised in goose fat, and you will still be as slinky as a well-hung fox. It's a religious event, like communion: eat me, I am your lunch.

I did, and it was delicious too: potted shrimps, a tiny little teal with excellent muscle tone, scarcely cooked, oozing blood as though it were gravy, and half a bottle of claret. The other half, the other teal and the other shrimps were eaten by Robin Hanbury-Tenison, who was paying. He was Chief Executive of the British Field Sports Society. They could not have had a more charming advocate and we ranged happily all over the environ-ment, society, ethics and politics as though we were old friends. It is a curious fact that animal killers are usually nice and Mr Hanbury-Tenison was obviously one of those people who have a certain ease because they are able to countenance hunting and killing. Always trust a huntsman.

We talked about cruelty. Didn't he mind the suffering of the fox as it was chased? Wasn't that unkind? There were two things about that, he said. A fox, or a hare come to that, either escaped a hunt or was killed. A hunt never wounded an animal which then crept away to die in pain. In that sense hunting was quite different from, and better than, shooting. Any number of hunts, particularly harrier packs, pick up animals that have been wounded with guns.

Yes, but what about the chase itself? It is the drawn-out threat of death that many people find most difficult to stomach. That, he said, swig of claret, lovely smile, is the second thing, and where they are wrong. He then told a story about a rat in a cage with a snake. His brother kept a snake in the Caribbean. It needed its daily live rat and every morning this is what happened. The snake is lying curled in the corner of the cage. He twirls his hand beside the bread rolls to demonstrate the curled snake. The rat is popped in at the other end of the cage. Other hand pops in and sits neatly next to the bottle of wine. The snake then thinks, 'Ah, breakfast.' Snake hand slithers along the table. Rat thinks, 'Here I am in a cage with a snake. No problem. I can deal with this in a hop and a skip.' Rat hand hops and skips over the table, easily and delightfully escaping the open mouth of the snake hand arching towards it in its gyrations.

The rat is not, as we might imagine it would be, because we ourselves might be, in terrified paralysis in the corner waiting for death to come. Not at all. This situation is only a slightly heightened version of everyday life for the rat. The rat is always in 'dynamic tension' with its environment. It is always thinking either 'Oh heavens, that's going to kill me' or 'Oh good, I think I can kill that.' That is the substance of rat life and its busy little brightness is a product of it.

So the snake stays slithering and the rat keeps hopping while this explanation of its psychology is made. Rat is happy; snake is hungry. Then, suddenly, the snake hand, somehow fused at this point with the hugely open-eyed face of Mr Hanbury-Tenison the other side of a table which I hadn't realized was quite this small until now, rises up in a surge of hunter-gatherer energy and – glup – swallows the rat whole.

A quivering little pause as the disaster sinks in. 'That', he said, his hands human again now and back on the knife and fork, cutting another slice from the breast of teal, 'is exactly what it is like for the fox.' Such was the realism of the enactment, and the passionate conviction of the *mise-en-scène*, that anything I might have said in response became immediately redundant. Hunting was nothing to do with red-faced men on big-bottomed horses farting their way across the English countryside. It was the far more charming sight of Robin Hanbury-Tenison's left hand suddenly swallowing his right hand whole, the laws of nature in action.

Back home, away from the sluicing of claret and teal-blood, I retreated, as ever, into the privacy of a quiet and secret relationship with the place. I shrugged my Kipling coat on. I was coming to know this valley in the way a man knows the feel of his own palms, blindfold, easily, without drama. I trespassed everywhere, ignoring paths, climbing every fence, pushing across hedges, finding the ways through that the deer had made. There's nothing like trespass. I've done it all my life, invariably alone and most excitingly at night. I've climbed into the garden of a great house in the early hours of the morning, sidling past the Renaissance pavilions, brushing past borders where the moonlight has turned the dahlias blood-purple and the lawns into Caribbean-coloured ponds. I have climbed the outside walls of a castle further south in Sussex, late one evening, thinking it an empty ruin and only found, as I topped the parapet, hauling myself over the lichened stones, that they held inside them a beautiful, tile-hung farmhouse, with a rose garden wrapped around it. The Iceberg roses, filling the bailey with their huge white flowers, were unreal in the moonlight at the farthest, darkest corners, and apricot and peachy where the light from the windows fell on them. No daylight garden, no allowed-in garden, could have matched it.

Trespass is an aesthetic experience, exciting, addictive even, because it is the most revealing way of being in a place that I know. It makes you slight, careful and attentive. You cannot stroll as a trespasser; it is not a breezing-about, hands-in-pockets way of being in a place. Trespass strips comfort from the mind.

I have never poached other people's game but I imagine poaching as an even tighter sharpening of the senses, screwing you into the details of the moment and the place, restoring an alertness and exposure to your presence in the landscape which centuries of pastoral urbanity – the smooth attitude to general rural effects – has clogged and obscured. To feel the immediate pulse of things, to be forced to shove yourself into a hedge when a stranger comes around the corner, the spikes of the hawthorn dug into your shoulder, the leaves against the face – that is what trespass is good for, the tangibility of the trespassed-on world.

Looking east from Jim's Field at
Perch Hill over the woods, fields
and farms of the Sussex Weald.

Adam Nicolson in about 1990, just as his life was falling apart.

Sarah Raven at the same moment.

The transformation of the garden
begins, getting much worse
before it could ever get better.

Friends help to lay out the beginnings of
Sarah's cutting garden, with woven hazel
windbreaks and a semblance of order.

Ben and Rosie Nicolson in the newly
laid out garden with brick paths and
far-too-expensive new pots.

Chestnut stakes, with turned onion tops turned, painted to look like gondola posts, with blue plant labels marking out the rows: Sarah starts to evolve the Perch Hill style.

Donald, by far the longest living of all our ducks, sorts out the slugs in an early version of the springtime cutting garden.

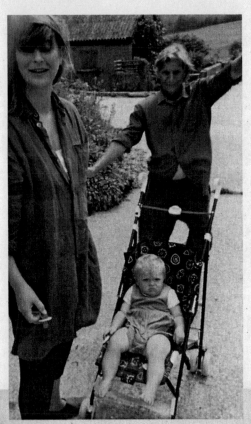

Sarah with Will Clark and a slightly disgruntled Rosie.

The new oast house emerges from a sea of mud, snow and rubbish.

Nipper Keeley, Sussex man par excellence, wood merchant, dog-breeder, furniture maker, wit and raconteur, gives Sarah a bunch of his supremely long-stemmed sweet peas.

Summer supper with Rosie, Adam,
Molly, Sarah and, half-visible,
Colonel Custard the dog.

Ken Weekes, born at Perch Hill, dairy
farmer at Perch Hill, by the 1990s tenant
of Perch Hill cottage, general guide and
advisor on every element of Perch Hill life.

Needless to say, being caught and ejected is horrible enough. I have had a Welsh farmer standing in front of me, arms crossed, resting his weight on the back foot in the way that Michelangelo's David does, his grey pork-pie hat tipped up off the brow, his cardigan waistcoat pouched around the stomach, while his three, circling, agile, grey-eyed dogs roared and screamed at me as I tried to make my way back off his private land on to a public path. Whenever I think of the word 'property', that is the picture that comes to mind.

The night-time wanderings here came to grip my imagination. Now and again, as it turned midnight and everyone else was going to bed, I would drink a glass of whisky, pull on a coat and go out of the back door. A moonlit night, the light so bright that it looked like a cold blue day. The sheep were coughing in the barn, kicking out their innards, hack, hack, hack, as if they'd smoked forty cigarettes since breakfast. It was dust in the hay that did it.

I set off into the fields. The open back of the hillside was washed in blue light. The further corners of the fields, where they dropped towards the creases of the streams, were bathed in deep blue-black shadow. It was the place I knew, transformed in the way that a fall of snow can rearrange the geography. But here it was time, rather than space, that seemed oddly altered. The moonlight created what looked like a form of summer. Its light cast the same sort of shadowy patches under the trees that you get in the summer sunshine. The moon was at the same height and at the same place in the sky as the sun had been during our lunchtime picnics in June. The pattern of light was the same; only the colour had changed. It was a strange and ghostly re-enactment.

And these winter nights it wasn't warmth and chatter but cold and quiet. A car on the Burwash road spread ripples of engine noise through the night. You could hear its gears changing as it came out down the long straight towards Burwash Weald and then moaned its way on up the hill towards Broad Oak and beyond. The silence pooled back in behind it, leaving only the sound of my walking, my breathing, the crunching of my feet on the frost.

Sometimes, and I don't know why, I began to feel frightened. Once, I thought I saw the shadow of two deer grazing in the Slip Field and I went down to get a closer look but they turned out to be nothing but clumps of rushes, masquerading in this single-tone light, this bleached ultra-violet black-and-white, blue-and-white light, as animals poised over the grasses. There was a slight creeping at the back of the neck, a crawling alertness in those vestigial hackles, raised by an ancient fear.

The frost made a crust on the grass under my feet, frozen on the surface and soft underneath. Where the moon reflected off it, the ice made pinpricks of light as sharp as stars in the grey-blue grass. The whole field was the colour of eyes but spangled with these lights. I didn't want to go into the wood. If I hadn't wanted to find out what it was like in the wood in the middle of the night, I wouldn't have gone in. I looked in at the deep black shadows in there and thought, no, I won't. But that was giving up too easily. I had to know what it was like and I went in. I hated it. The inside of a wood at night is a darkened, complex, crabby place, full of the flick and twist of shadow, a place of misinterpretation, of reading the neutral and the inno-cent as strange and threatening. All those trees look so human. They look too like living things for comfort, they surround you and they outnumber you. There is a sense that they are all at the back of you, those outstretched arms, those leaning limbs, that overarching command of the place they have. That is the sort of idea which, once it has made a bridgehead in your mind, won't be dislodged. The fear moves in and colonizes you. Even as it was happening, I could not believe it was happening. Never would I have believed that these woods, our woods, could have scared me, but they did. What is it, this deep, black fear of an uncontrollable landscape? Modern neurosis or ancient genetic memory? Is it basic or is it acquired, something essential to our own natures or a symptom of distance from the natural world?

I wanted to be out the far side of it and I ran crouched through the moon-shadows, through the snatching brambles, snapping the twiggy elder branches as I passed, until I was out again in the wide, grassy, controllable ease of our far big field called Great Flemings. Panic at shadows? Surely I should be beyond that now? After-twinges gripped the back of my neck and a dog yapped somewhere down at the bottom of King's Hill, a mile away, a persistent, repetitive, dreadful barking as I caught my breath.

I sat down in the field and looked out at the farms on the far side of the valley. I knew from the hedge shapes, the dark lines of the hedgerows laid across the grey of the fields, where the farmhouses should have been, but they had merged into the dark. There were no lights on over there. A long way over to the west a jet was making its way into Gatwick and there was a sprinkling of lights further north, fewer lights on the earth than stars in the sky. It seemed as if everything local had dissolved away, leaving nothing but the dark and the calling of ewe to ewe, from points up and down the valley.

And then into that, the wholeness of that, tore the one sound in the English night which slashes bigger holes than any other: a fox, the scream-howling of the fox in the wood, a noise gash in the rest of it: *yeaow, yeaow, yeaow,* a ripping of cloth with claws, *yeaow, yeaow, yeaow,* as screamingly basic a sound as England makes.

Neighbours with the Dead

THERE IS a small lane that runs over our land. It is, as I see it anyway, a beautiful place, with a hard stony surface where the wheels go and a grass strip between them, hedges 8 feet high on either side which, at the far end, lean over to meet at the top, so that in summer it is a green and sun-splashed tunnel, full of broken shadow and leopard-skin light. Some 200 yards after it has left the road, it finally curls round a little into the wood like the warping of a plank that has been left out in the rain.

I love this lane, for the roses and honeysuckle that hedge it in early summer, for its privacy, its following of an old line off our hill through the wood and on down to the valley at Bateman's. If I were designing a landscape, it would be full of tracks like this that can't be seen from the outside, trenched inside their own hedges, eventually curving away into apparent nothings. Landscape beauty, you might even say, consists of slight disappearances like this, not abrupt corners or sudden conclusions but turnings in as subtle as the spin on a ping-pong ball as it crosses the net. That slow, slight curve appears again and again in descriptions of landscape that are intended to catch the moment of perfection. In Philip Larkin's poem 'The Whitsun Weddings', he describes a train journey from Hull to London, and the air of warm, early summer completeness in it is set up by that one resonant word: 'A slow and stopping curve southwards we kept'.

There is a taut balance in a shallow curve which other shapes could never match. When Ben Nicholson was persuaded to discuss his own painting, he would do so, if at all, not in terms of the influences on him or the meaning of what he was doing, but of spin and dancing, the perfect execution of a turning step, the alert, controlled, coherent energy of it. Or there is Edward Thomas dedicating his book *The Icknield Way* to his friend Harry Hooton, saying that as far as walking is concerned, or even *being*, if it comes to that, 'the end is in the means – in the sight of that beautiful long straight line of the Downs in which a curve is latent . . .'

From the beginning I loved our lane because it was an unconsciously beautiful thing, in its shape and in the materials of which it was made. But there was a problem. The lane belonged to us but was the only means of access to another house. We owned the lane but we hardly used it. Dave Fieldwick, who had his sheep on some of our fields that winter – I was charging him 30p a week a ewe for the winter grass – used the lane once a day to check if they were all right. Otherwise the only users were our neighbour, Shirley Ellman, the postman delivering her mail, the electricity man reading her meter, her friends coming for a dinner party or her children on a visit at the weekend, and so on.

Things became tense between us. The potholes grew steadily worse. One particularly big one filled up with water and the children and I spent a happy afternoon trying to sail plank-and-hankie boats across it but it wasn't quite deep enough and they kept grounding. This Perch Hill Round Pond was not seen as an amenity by Shirley and she decided to do something about it. The first thing I knew, going for a walk one afternoon, was the sight of a large yellow lorry, two men and a rolling machine laying out shiny new tarmac on the surface of the lane. By the time I got there, they had covered 50 yards of the lane with the smart new blackstuff. 'No, no, no, no, no, no, no,' I shouted.

The tarmac looked – was – horrible, a suburban slick, removing meaning from the place it coated, making the lane banal and ugly, no longer a place but a passageway. I had rather surprised myself by the intensity of my reaction. Could such things really matter so much?

The following evening I had a very civilized meeting with Shirley. I explained it was my lane and she shouldn't do things like that to something that wasn't hers. She said the ruts were atrocious and the suspensions of their various cars were suffering and that's why they had put the tarmac on. I said the tarmac destroyed the meaning of the place. She said what on earth did I mean. I did not mention Philip Larkin, Ben Nicholson, Edward Thomas or 'straight lines in which a curve is latent' but I did use the word 'suburbanization'. She said she thought she was improving the lane and that I as the landlord should be grateful. I said I wasn't grateful to have what I owned altered by someone else without even so much as a by-your-leave, and that some people in my position would have had the tarmac dug up and the bill sent to her. Surely, I then said, calming, calming, we could come to some mutually satisfactory arrangement? She agreed, surely we could, but what was I proposing? I proposed that we should pay for the upkeep of the lane in proportion to the amount we used it, which we should work out, that I would get a quote from someone to bring it up to a beautiful, untarmacky, usable condition and we should go from there. She thought that sounded OK and that was how we left it.

It could only be an interim position. There were two irreconcilables in conflict here. I wanted the lane to look and be rural, of the place, not alien to it. My neighbours wanted it to be usable, reliable and undestructive of their cars. I wanted a stone track, which would inevitably be eroded by the sort of traffic a modern household generates; they wanted a tarmac drive, which would not. What could be done?

In part I felt, secretly, without saying so, that I was being unreasonable. Why not allow these people to make their own runway across your land? But in another way I felt, symbolically, that over this I had to dig my toes in. What else were we here for but to nurture precisely the wrinkliness in the landscape which that lane represented?

The poetry is in the particularity, and the particularities of particular places are not single elements dropped on a desert floor, but webbed, interfolded with each other, making a multi-dimensional grid over the extent of place and the depth of time. Anything with clarity will be clumsy in the face of that and, when dealing with webs, clumsiness is the one thing that has to be avoided.

But we could come to no agreement. The untended lane remained there for months on end, rutty, their cars lurching down it, the fifty yards or so of tarmac at one end pristine, the rest in decay.

The situation was made worse by another problem. Our water system was arranged so that I had to pay Shirley's water bill and she would reimburse me for it. This became entangled with the lane issue. She refused to pay her share of the bill. We threatened to take her to court. Only months later did we discover that there was a leak in the pipe between her meter and her house. She had never seen the bulk of the water the meter said she was using. These were the ingredients of a bitter relationship between us. We passed each other in our cars on the way to and from the village: stiff smiles or bleak indifference through the windscreens. Occasional curt phone calls. I started to exclude her presence from my own picture of the place.

It was not as though the rest of Perch Hill was in a very good state that winter. Sarah and I, as people do in these situations, had embarked on radical reconstruction of house and garden. As if deep in analysis, things at the farm were still on the downward slide. We were still deconstructing its personality, aiming steadily for the black chaotic pit from which, and only from which, the rebuilding could begin. If there were mental hospitals for landscapes, Perch Hill would have been sectioned that February. Poor, broken, misused, fragile thing.

When people, usually of the older generation, came to lunch here on a Sunday there was always a moment of awkwardness, a rather uncomfortable hiatus in the social flow, when they would usually have said how beautiful the house/garden/surroundings were, how exquisitely lucky/clever we were to have found such a perfect hideaway and how on earth did we ever stumble on it in the first place?

I started getting used to the approach of this pre-lunch crisis and could see our guests silently struggling with the problem. I pushed peanuts and cheese straws at them but it wouldn't go away. What possible compliment could they make when half of one of the kitchen walls was missing, awaiting a replacement fireplace and held up by a rusty old acro prop? When the terrace outside the kitchen was a mixture of a trench dug by mistake for a wall we would never build, incipient weeds and a zigzag brick pattern which looked nice from an upstairs window but twisted your ankle if you walked on it? When, on the other side of the building, the groundworks for a new septic tank and reed-bed sewage system – very green, very Highgrove – had made what used to be a perfectly acceptable lawn into a mud stew. My sons described it as sad. ('What does sad mean, William?' 'Sad is a punky word for stupid.') When what was laughably called the scullery was a muddy mixture of plastic tree guards, sheep's veterinary equipment, unmended hoovers and put-it-there-for-the-moment junk?

What could they say? 'I see you've taken quite a bit on,' was the usual version. 'How long have you been here?' was another, a little more sly – or perhaps shy. Only the brave could admit that 'Every time I come to this house it seems to get less built.' And only they were right.

But Sarah and I had something in mind. If we were thinking of becoming high-profile management gurus, we would have called it 'The Knowledge'. The Knowledge was what kept us going through the mud, like the differential lock on a Land Rover. The Knowledge was what, with time, and the drip drip of in-dripping finances, we would make of this place. It would be whole and it would be good. It would have the air of what might be called inherent coherence.

Inherent coherence: it is something you recognize every time you come across it by chance, driving down a lane or turning a corner to a small group of buildings and trees, or often in an unmetalled track, where a sense of well-made purpose suffuses the little patch of landscape. These are places where the essentials are what give them character, where the meaning or the style of the place is not pasted on from the outside but emerges from within, where there is no sense of hypocrisy, no lying, smiling front to a cold and cynical core.

It would be good. It would have to be good. One week, later that winter, I saw a picture of what Perch Hill might one day become. I was in the University Library in Cambridge. Falling asleep over what I was meant to be reading, I decided to see if Perch Hill Farm featured in that building's vast subconscious depths. There was nothing in the computer catalogue, but that was probably too much to expect, so I went to the Map Room. Here on giant green tables a serious man, writing the history of Bechuanaland in the late 1890s, was poring over garish maps of mineral deposits and catchment areas; a woman in an Inca-style cardigan was analysing the tide streams in Scapa Flow. I asked for Perch Hill Farm. 'Certainly,' the map librarian said. 'Just fill in the form.' She disappeared for a minute or two while I kicked my heels and she returned with one heavily and precisely folded piece of paper.

She left me to it and carefully I unfolded the sheet. It was large, perhaps 3 feet by 2, and had a clean and precise air to it, as if freshly laundered. There was even the smell, in its inner sections, of newness and ink. But it was far from new. This map was part of the great twenty-five inches to the mile series made in the second half of the nineteeth century. This particular sheet was produced in 1898. I don't think anyone had looked at it since it was made.

At my own giant green table, I pored over the map of home. The farm just about filled the sheet. The other people may have been thinking about, or analysing, or drawing conclusions from the maps in front of them. I was not. I was in bed with my map, loving every inch of it, drinking it up, reading the reality of hedge-bend, gateway, wood-corner and stream-turn, surveyed so exactly, drawn so carefully, displayed so perfectly in front of me. This map series, which marks individual trees in hedges and names every field, which, if laid out for the whole country, would stretch 200 yards from the Lizard to the Cheviots, scarcely less from Southwold to St David's, is probably the greatest map ever made.

I looked at my sheet, one tessera of a stadium-size mosaic, and in it saw the state of perfection, described in a fortnight's work in the spring of 1898: the hop garden in Hollow Flemings, no longer there; the small wood that cut in two the big field known as Great Flemings, no longer there; the three hedges that made small compartments of the other big hay meadow, the Way Field, none of them now there; the little wood dividing Target from Cottage Field, marked now only by a bank and a single oak; the orchard in the Cottage Field, of which one fruit-less plum tree remains.

Here, in the Map Room, surrounded by the nearly audible sound of the collective Cambridge brain ticking, I saw something else: our farm in its rich, divided wholeness, the picture a century ago, the agenda for the next 40 years. A wide destructive gash lay between me and the moment the map had been made. The small-scale agriculture of the Weald, dominated by 90–100-acre mixed farms cut out of the wood, was still more or less intact in the 1890s. It was not a world any longer of self-sufficient yeomanry. That had long gone. And the presence on nearly every farm of the hop garden and the oast-house in which the hops were dried was a sign that the tendrils of the London market, and its thirst for beer, had already reached deep into this stretch of country via the railways. There were already London commuters in late Victorian Sussex. Just down the road was the first battery chicken farm in this part of the world. This was no prelapsarian paradise. But the shape of the farm then, with its closely divided fields, its well-maintained coppices, its imposition of order but application of care, were all signs that like other poorish Wealden farms this one would only sustain a family if maintained like a properly screw-tightened, lubricated engine.

Much of that had gone in the intervening years. Hedges were taken out, fields enlarged, woods abandoned, grassland re-sown and dosed with weedkiller and nitrogen from a bag, big old oak trees felled to make way for bigger sheds and barns with railway sleeper walls and asbestos roofs so that a larger herd could be kept here, more money made and the land driven down essentially unsustainable paths. The very market-based thinking that was doing this to Perch Hill would mean in the end that commercial farming here could not survive. In the 1940s a hideous new cow shed was made out of the cheapest and nastiest of bricks (from Bedfordshire); a bull pen was constructed out of crudely poured concrete. In the drive for productivity, almost any sense of here was nearly abandoned in the 20th century. It survived only in those remote corners where the spirit of enterprise couldn't quite reach. It is a strange fact that Perch Hill Farm, like England or Europe, had its core and its periphery, its busy, transforming centre and its outlying limbs where a longer, calmer rhythm continued to be at work. Those remoter parts of the farm became the reservoir of the past, and even of value, from which we could draw for the future. That was why this map was so beautiful to me. It showed Perch Hill before any of the destructions occurred. It was a picture of the place as I hoped it would one day become. Here it was as it might yet be.

I rushed home. I shrugged on this place like an old dufflecoat: an arm in each sleeve, a quick flick of the shoulders and the thing thudded on, into place, as if it had been hanging on the coat-hook pre-formed. I rolled outside, coat on, boots on, hands in pockets against the finger-nipping wind. The ducks were beside the old cow shed on the pond, whose level had risen higher than at any time since we'd been here. They remained fearful, but that was as it should be, their own fox-proofing. The chickens ignored me, another sign of health, busy in the hay beside the barn door.

But all this was part of our still-to-be-sorted farmyard. What I wanted was the fields, the eiderdown expanse of the bed of home. I could feel myself wriggling down into it as I entered the first of them, the yellow-green of the sheep-nibbled grass as welcoming as any nursery ever could have been. Was this what the fox felt when he entered his earth, that surrounding of yourself with a place that seems to be an extension of yourself, so that there is no sharp transition at the skin?

As I went down through the fields, looking at the beautiful reduced colours of the winter wood – the blackberry purple of the hazel fronds when seen against the sky, the yellowness of the willow in bud, the chocolate blackness of the dead bracken where it lay sodden and rotting at the field margins, the grey-eyed green of the lichen spots on the birch trunks, curling up at the edges into a sort of lettucey thickness – I could sense this whole farm webbed with the different overlying territories of the creatures that inhabit it.

It took going away to notice this, to look beyond the details of daily business. The multiple, interlocking mesh of this place became apparent. I realized that my own sensation of rooted, inner belonging could only be akin to the territorial sense of the foxes, the badgers, the deer, the wrens, the robins, the kestrels, the pheasants, even the bees or wasps or those chalky blue butterflies that had flickered above the surface of the summer grasses on the limy lower ground of the Slip Field, whose descendants must now be holed up somewhere near by, waiting for the warmth to return.

There is so much said and written about biological community, but never before had I felt myself to be part of one. Nor had I recognized that what I had always thought of as a higher human faculty, this identification of self with place, was nothing more than an animal faculty for location, a means of placing oneself with precision and care in the landscape, an essential adaptive tool for survival. I couldn't actually see the other animals doing it, but I knew they were there. I sometimes caught a badger in the headlights on the lane, so much fatter and piggier than you might expect, trotting up into the brackeny laneside banks. We buried a large vixen that was lying dead and fresh on the side of the road just before Christmas, the wind ruffling up the white hairs on her throat as she lay on the grass while I dug the hole. The deer would melt in and out of the wood in the first hours of daylight.

Seen only by chance, these animals were a constant presence. I felt no strangeness from them. They left their marks. On a dewy morning, the fields were criss-crossed with their tracks, where their bellies had dragged through the wet grass. Where the deer jumped the fences, the earth was punched away. In the Way Shaw, down on the edge of our land where the ground falls sharply towards the river, the deer had pushed out nests for themselves among the birches and bracken of the heathy woodland. The owl that sat in the giant beech at the corner of the shaw I had often looked for and often heard but never seen. It was as moulded to the place as cheese melted into the crust of a pie. But I now think, perhaps, that to want to see it, to get a visual grab on it, would only be trophy-hunting. The truer knowledge of it is to hear its unseen presence in the trees. You don't have to open up someone's chest to know they have a heart.

My walk around the fields slowed down. I had started off briskly enough, seeing that this gate was working properly now, that stile needed replacing, checking the sheep were all right, noticing that one was limping, but as the walk went on, the progress was tripped up by the details. A woodcock suddenly flipped up and twisted away in front of me a shadow-bird, part of the wood-floor on the wing. I noticed an oak tree still covered in its brown autumn leaves, a blackthorn still laden with sloes that were burst and crusted as though cooked under a slow grill, the way the red hawthorn berries, seen against the blue blackness of the woodland, merged the two colours so that the landscape itself turned purple. Slower and slower, I noticed the ear-like fungi growing out of the elder trunks quivering in the breeze like blown flesh, the spindle berries the colour of a liquorice allsort, just going over, a wren, incredibly, not hopping but flying through the branches of a hawthorn hedge. Perhaps the slowing down was this: the reassertion in me of animal life, the reading of familiar signals, triangulating the map of the known.

I realized that something of the secret here consisted of burrowing in; and that the unsatisfactoriness of modern life was based on a wrong geometry. We tend to graze, browse, pause and surf, glancing not staring, acquiring the bought article with a card or some numbers typed on to a screen. There is no digging in any of that, no sweat, no engaging with the understorey, the lower strata on which all the bright desirables must rely. The one thing that people in this sliding, gliding, acquiring life all hunger after is *depth*, the going in, the plunge. Perch Hill for me was exactly that kind of going in, a sounding-bell in which I could feel life not as a thin, two-dimensional grid but as an encompassing universe. Depth and thickness were at the heart of it.

Above all, I poked around our streams, but to call them that is to flatter them. One drops westwards from the house and one east. They are trickly, weedy things, dry all summer, with a proper gush only after heavy rain. You could never have a boat-race with sticks on them. But then, saying that, I'm aware only of disloyalty. They are, in their tiny way, beautiful things. The slight winter flows make small waterfalls where the stream exposes a bed of the underlying stone, followed by small pebbly pools no more than 18 inches or 2 feet across. Hart's-tongue ferns fringe these places. Come the spring, one stream will be bedded in garlic, the other in bluebells. Both of them are deep in trees, one in the edge of the wood, Dallington Forest, which stretches westwards without a break for 3 miles from here, the other in its own little strip of trees, somewhere between a massively overgrown hedge and a sliver of wood.

There was a hidden aspect to these streams which only stole up on me very slowly. We had been here for months before I realized. They seemed in themselves such inconsiderable things, transitory, unimportant parts of a landscape dominated by field and tree. But as I realized late that winter, they were, to use the phrase of a French historian, Alfred du Cayla, describing the streams and rivers in the landscape, 'les lignes maîtresses du terroir', the mistress lines of the territory. I love that phrase. Our little streams embody it. They are a governing presence. Whenever it was that the boundary between the parishes of Brightling and Burwash was drawn here, certainly before the Norman Conquest, perhaps centuries before that because the first *written* reference to Brightling was made in AD 732, they used these two streams as their line, climing up their beds from the valley to the high ground where the farmhouse now is and then down the other side dropping steeply to the river, three hundred feet below us.

The hedges that run along them, as you would hope for from a parish boundary, are enormously rich in tree species, 12 of them: crab apple, spindle, hawthorn, hazel, blackthorn, oak, ash, field maple, rose, alder, willow and elder all grow there. There is a widely accepted but slightly unreliable rule that for every woody species you can find in a hedge, you can add a century to its age. The large amount of surrounding wood probably boosts the number of species in all hedges here. Nevertheless, if the rule is at least some indication, then those hedges, that acknowledgement of the tiny stream as a mark in the landscape, are at least a thousand years old, maybe more. The hedge date would just about match the date of the first written record.

The history hunger built inside me. Nothing of any detail could be rescued from the Dark Ages, but later, when this farm was cut from the wood, there should be something there. What was it like, when this place first emerged as a place and what, above all, was the man like who first made it? I looked for hints of what he did, the marks he left, sifting through them in the landscape in a strangely possessive and jealous way. I wanted to know it all, in detail. It felt at times like digging through the waste-paper baskets for someone's leavings, their inadvertent signs, the give-away gestures.

But he had covered his tracks, or had his tracks covered for him, well enough. What evidence there was in this microdot of England is slight and fragmentary. You could not say that silence hangs over the beginnings of this farm, about four hundred years ago. More, the fuzz of what has happened since obscures the shapes and muffles the signals.

Even so, there is something about this searching for another, this listening out for the lingering presence of someone else's acts and motivations, which is curiously intimate. I am not saying that this is a haunted place, only that understanding can persist across time, that sometimes, perhaps because I have come to know the wrinkle of field and wood here as closely as he would have known them, I can imagine, at least, that there is a sense of understanding between me and the man who made this farm. His person is in the fine grain of here. In some ways, this place is his mind.

In the East Sussex County Record Office in Lewes, I made my way back through the documents. In the 19th century the farm had belonged to the Fuller estate at Brightling, originally brought together by John Fuller, known to history as Mad Jack, a rabid Tory Member of Parliament, sugar-and-iron millionaire, folly-building eccentric, patron of Turner, buried in Brightling churchyard in a pyramid where he is said, erroneously but appropriately, to be sitting down to a final gargantuan dinner at a cast-iron table.

At Perch Hill in 1832, a man called Edward Goldsmith was his tenant, and stretching back through the 18th century was a string of tenants, none of them more than a name. In the late 1770s, there's a Mr Carter, in 1771 a Mr Harrison, in 1724 a Thomas Noakes, in 1719 a John Baker. In 1711, a John Taylor lives at 'Pearchill' and he is already there in 1694. The farm was then valued at £9, the poorest farm in the parish. Perryman's, the Wrenns' farm on the other side of Leggett's Wood, was valued at £19 and Park Farm, down by Bateman's on the good alluvial land, at £34.

Here and there in the sale documents, there is slightly richer description. When Jack Fuller first bought it in 1820, there was a house (ours), a cottage (probably Shirley Ellman's), a barn (on the site of ours) and a coachhouse, a stable, 20 acres of wood, 40 of meadow and 40 of pasture – almost exactly what is here now.

All of the eighteenth century farmers were tenants. The owners of Perch Hill were a variety of professionals: a London lawyer, a 'chirurgeon' of Rye, an apothecary of Tonbridge, a tailor of Wittersham in the Isle of Oxney, and other men whose status fluctuates in different documents between 'yeoman' and 'gent'. The ancient picture of a man and his family living on his own smallholding had not been the case at Perch Hill since the end of the 17th century.

Before then, what is called 'Perchfield house' is distinguished from 'Perchfield lands' and from another, historically separate holding called 'Flemmings'. These intriguing terms must mean, first, that the fields at Perch Hill were in use and known as that before there was a house there. Two of our fields running down to the Dudwell are still called 'Great Flemings' and 'Hollow Flemings', probably belonging at some stage to a man called Fleming. It was a common surname in England throughout the Middle Ages.

That name remains part of the Perch Hill landscape as the documents drive back into the seventeenth century. In July 1650, a shopkeeper in Dallington, John Goodman, sold to Henry Goldsmith, gent of Burwash, 'a house & pieces of land called Perchfields and Flemmings in Burwash'. I thought for a long time this was as far as I was going to get, but the archivists in Lewes have pushed the story back still further. In 1603 a yeoman in Burwash called Henry Weston gave Perch Hill to his son Thomas on the day of his marriage to a girl called Agnes. The father kept 'other lands called Flemmings' for himself but gave Thomas and Agnes 'a messuage called Perchhowse and a barn and lands called Perchfieldes, to the said messuage adjoining and belonging'. There is no doubt these buildings were our buildings: the document describes their precise relationship to Willingford Lane and to the lane that goes past Shirley Ellman's house and down to Bateman's, then called 'Perch Lane'.

But there is a further surprise. To the description of the house and its situation, the document adds 'with garden belonging'. An Elizabethan garden at Perch Hill! What a life-enhancing discovery that was! There would have been vegetables and a pig or two and a beehive but maybe rosemary, pansies, marigolds, columbines, violets and sweet Williams as well. The Elizabethans loved roses. Ophelia and other Elizabethan heroines mention all these flowers. But even that was not quite the beginning. The deed selling this place in 1581 had survived in the papers of the Fuller estate. 'Pearchefield' and 'Flemmyng' were already bound together, with only 27 acres in all – the smallest of Wealden smallholdings – but with 'tenements' mentioned in the deed, meaning there were buildings here. Perhaps this was the Ur-moment, the point at which the first oak frames were erected here, the first thatch laid (the marks of the ropes binding the thatch to the rafters can still be seen in the attic), the first fires lit, the first nights slept, the first children conceived under the roofs of Perch Hill. It felt like yesterday, an embracing of the past.

Only one further document pushes this time-line deeper into the medieval Weald. On 15 November 1419, the seventh year of the reign of Henry V, a Burwash man gave his son Richard 'a piece of land called Swetyngecroft'. That name had already dropped from view four hundred years ago but Weald historians have established, from tracing the holdings of the neighbours mentioned in the grant, that Swetyngecroft was Perch Hill itself. There were no buildings here six hundred years ago, but at least an identity and an ownership. The name itself has some obscure hints in it. 'Croft' is clear – in Old English a small enclosed field. 'Swetynge', which was a medieval surname, could mean either 'sweating' or 'sweet little thing'. So what was the Perch Hill croft: Mr Swetynge's field; or a sweet, darling little place; or one demanding hard physical labour; or one maybe that sweated itself, oozing water, its springs seeping from the clay underlayers in rushy clumps and reedy patches? It could be any or all of them.

The story confirms something. This farm is right out on the edge, the worst land where the border between the parishes runs, which no one would have bothered with unless forced to, no more than marginalia, the periphery not the main. One of our fields is called, rather enigmatically, Toyland, and the little wood next to it the Toyland Shaw. But that, I came to realize, is a mishearing of something said in widish Sussex. It wasn't Toyland but Tyeland, part of the tye land, an Old English word meaning a common, of so little intrinsic value that it could be left to supply the common needs. And that too lies at the heart of the farm's name. Perch Hill has nothing to do with fish, nor being perched up high, but simply the hill where you would go to cut your perches, the old word for a good stout stick or pole. A parrot's perch and a perch as a unit of measurement are both more specialized derivatives of that older and more general meaning. 'The tame Hoppe,' Henry Lyte wrote in his *Niewe Herball or Historie of Plantes*, published in 1578 as this farm was coming into being, 'windeth it selfe about poles and perches.' So this is where we live: Stick Hill, part of the medieval commons, out on the edge and stuck away, well and truly in the sticks.

All this was inching us towards my man. Knowing this edginess of the place, you could start to smell the quality of the man who first decided to live here. He was not part of the establishment, nor in line for some major inheritance which he only had to outlive his father to receive. He was, perhaps, a younger son who needed to make his way. There would have been those dissuading him but he wouldn't listen. What other choice did he have? The lottery of the cities where you were more likely to die young than do well? Or paid employment somewhere, a ceiling imposed on your prospects and your life? Neither would do. Here, on the boundaries of the parish, there was rough, poor land which could, with work, be turned to use. It had not been wildwood for a long time but no one had ever called it home. There were more people about than anyone could remember. It was time to do something new and here in a small way was a new world to conquer. That was Perch Hill: a fragment of America embedded in the wood. Its creator was a colonist.

Who knows how it went to start with? You can only look to the fields. Those around the house, including the Toyland, have the wriggliest of boundaries. They can only have been quickly cut, fitting around existing lumps or immovably big trees and then ossifying in position. The Long Field, which slivers down between the stream and the Middle Shaw, is not now as long as some of the others and so its name must record an earlier state, when it was indeed the long field in a group of tiny wood-cut enclosures. The Target Field beyond it is almost as round as the name might suggest and that too makes it feel early. Where they meet the wood on all sides, an old bank and ditch, dug by our man, still marks the boundary between them, the crucial separation of stock from growing trees. Only on the good corn ground on Beech Meadow would he have planted his first crops and there, in fact, is the mark of many centuries of ploughing: a lynchet, a thick belt of soil built up against the downhill hedge.

All the ingredients of a self-contained world were here – shelter, wood, grass, water and corn. For almost a year I had been reading these things nearly every day, scanning them for the hints they might offer, what they might tell me of the lives that had shaped them and the lives they had shaped in return. There was a community here of another kind. We had become neighbours with the dead.

But what about the life of those first men here? Search as you might, there is no autobiographical account in English of a farmer's life in 16th-century England. Diaries were not kept and destinies went unrecorded. But I came across someone else that winter who, for the time, had created a miracle of self-description. He was a Frenchman, a Norman, and not quite of the right class – he might have lived in Bateman's rather than Perch Hill – but his mentality seemed to match these fields and woods.

Gilles Picot de Gouberville was the *seigneur* of the small village of Mesnil au Val in Normandy in the middle of the 16th century. In every aspect of his life, except one, he was completely ordinary, conducting an existence which in its enclosure and untroubled stability was utterly typical of his time. One thing only marks him out. Between 1549 and 1562 – that is between the ages of twenty-eight and forty-one – Gilles de Gouberville kept a journal every day, writing between ten and thirty lines, usually in the evening. He never missed a day, whether ill, tired or busy, and he never revised what he wrote. His diary, of which a couple of studies have been published, is the only surviving, non-literary, unsentimentalized and undistorted depiction of the life of a rural village in 16th-century Europe. Nothing in English can match it for the pristine, unmythologized quality of the life it depicts. This is rural existence in Europe before the urban began to distort it, a written portrait of an unwritten world.

Again and again, Gilles says simply, 'Spent all day at home,' 'Here at home,' 'Here in the house.' His French editor doggedly counted the phrase 'I did not leave the house' 3,310 times in the thirteen years, or on more than two-thirds of the days he records.

Gouberville has only an approximate idea of time. Things happen 'at some time in the morning', 'towards the end of the day', 'a little later on'. The clock had yet to become the master of the working life. Gilles had one, but he kept it upstairs so few people saw it. When someone asked him for it, he gave it away.

Gilles plods on day by day in the long rhythm of rural existence. The hay has to be in before the rain, the wolves driven from the flocks, the boars from the oats, the herons and wood-pigeons into the nets. Everything is externalized, as though the conception of an internal private existence has not yet been invented. There is no intimacy, no self-consciousness beyond the fact of the journal itself, no feelings expressed, no sorrow or pain. The account is consistently modest and opaque. Although Gilles's life is surrounded and enmeshed with those in the village, he has no wife and no legitimate children. There are hints that he has an affair and possibly a child with a woman in the village, but that is not clear, and few women appear by name.

There is, however, a profound social fluidity to the way he lives. The farmhouse door is no barrier, but a thoroughly permeable membrane through which the village ebbs and flows. The people of Mesnil could visit Gilles even in his bedroom and he, with no great ceremony, sometimes came down to meet them still wearing his nightshirt, standing at the kitchen door or sitting at the kitchen table with them in front of the fire. On occasions, he wakes up to find villagers standing patiently by his curtained bedside, waiting for him to open his eyes so that he can tell them what he wants done. One day, in the summer of 1556, before he gets up, he buys a couple of pigs from a neighbour while still in bed.

His stone-flagged kitchen is the focus of village life. It is the only constantly warm room. In winter he gets dressed in front of the fire. That is also where he makes and receives any payments that are due. He occasionally sleeps there too and when convalescing, after several days in bed upstairs, he does so wrapped up in the warmth of the kitchen, where his neighbours come in for hot drinks or dinner.

If you read general histories of sixteenth century Europe, the air is agonized, fraught with crisis, tensed with growing short-ages of food and land, with the sense of repression and the expanding state, focused on the terrors and adrenalin of religious war and imperial ventures. But here, in the virtually pre-literate world of the Cotentin peninsula, some sense of wholeness prevails. This small-scale gentleman lives embedded in the milieu of his everyday companions. His half-brother and -sister live in the house with him, his father's illegitimate children, occupying a lesser but still intimate place in his life. Alongside them are his right-hand man, Cantepye, from whom Gilles is rarely sepa-rated, Arnoul, the secretary of the *manoir*, and La Joye, the lackey.

Outside this central knot, a wider circle revolves. The men who work on the tasks of farming life, the agricultural labourers, are named over and over again in the journal. They are not an undifferentiated mass, but individuals, known for who they are. Every summer the same reapers return for the harvest when more hands are needed, the same roofers, carpenters and masons return when repairs are needed to the mill or the roof of the *manoir* itself. A cooper lives and works in Mesnil and a black-smith, Henri Feullie, who shoes the horses, puts bolts on the doors and makes hooks and sickles. When a new cart is needed it has to be ordered from further away, from the cartmaker Clément Ingouf who lives in the village of Montaigu. Every year for a few days, Thomas Girard, the travelling tailor, comes to stay at Mesnil, equipped with his cloth, his tape, his scissors, his needle and thread, to measure up Gilles and his household for suits of new clothes.

Gilles is clearly friends with people from a wide social back-ground. There is no stiffness or formality in the way he deals with them. He is a particular intimate of a peasant farmer, Thomas Drouet, the sort of man who might have lived where I live now. Gilles is godfather to one of Drouet's daughters. Gilles and Drouet often work together in the tree nursery, the most treasured part of the estate, where the *seigneur* carefully tends and grafts his young fruit trees. When Gilles is ill, Drouet spends the night in his lord's bedroom to look after him; when Drouet himself has an attack of gout, Gilles visits him regularly at home.

Of course, Gouberville's odd and inexplicable journal marks the ending of the world it describes. On the cusp of self-consciousness, it hints, for all its descriptions of social cohesion, at the profound isolation of the individual, the diarist alone with his diary, which is the defining mark of the urban civilization that was to come. People moan about suburbanization of the countryside now; its first tendrils were already apparent in the Cotentin in the 1550s.

Could I translate Gilles to the Sussex Weald? Or perhaps Drouet? Would my Drouet here have spent time with the *seigneur* up in Brightling, tending to him, easy with him, living that webbed life of which I too had discovered the remnants here now in the late twentieth century? It is impossible to say, too much of a guess.

I was nostalgic for Gilles's life, even if I knew he would have

lusted after mine. I could see us standing at either end of a
corridor 450 years long. We can hardly make each other out but
we stare hard at this semi-familiar creature who is staring hard
at us. I live an unimaginably cosmopolitan life compared with
his; he is unimaginably embedded in his world; he is continuous
with his own past; I feel only tenuously attached to mine; he
looks at the level of comfort in my life in the way that we read
of sultans and film stars; I look at his rootedness – he is like a
human carrot – and long for the world in which I could give
my clock away and could buy a pig in the morning when still
under the blankets.

In my life, these social connections were not inadvertent and
casual as they were for Gilles. I had to engineer them; I had to
acquire the web he knew as normality. And so, for example, that
winter I signed up for a course run by the local Farming and
Wildlife people. Book early, their flyer said, because places are
limited, so I did, before Christmas, and eagerly acquired the
necessary kit: leather gloves with gauntlet-type cuffs that run
back up the arm, a bill-hook, once the commonest of tools,
some long-handled loppers and a smallish axe.

I imagined the others coming on the course preparing themselves all over East Sussex. Then I rang the organizer to check
the details. 'Oh, yes, Mr Nicolson,' he said ominously. 'The
hedging course. Yes. You must have been reading my mind. I
was just on the point of ringing. It's been cancelled. You're the
only person who has applied.'

So much for the great revival of rural crafts. You spend your
life thinking you are part of some widespread socio-cultural
phenomenon only to find that everyone else has decided to pack
up early and go inside for a cup of coffee. So I got in touch with
a hedger off my own bat. Could he come here for a day? No
problem. A gentle, definite voice. He'd come Sunday and be
there at 8.30.

In a cartoon version of these things, Boots, as he asked me

to call him, would be a gnarled old oak-bole of a man, solitary as hedgers are meant to be, reluctant to make much of a speech out of things, more articulate with his hands than his words. Not Boots, despite the nickname. He is a words man, endlessly weaving talk into the practicalities of what he is doing, about his life as a smallholder and now a teacher at an agricultural college at the other end of Sussex, where he tells students how to drive tractors, use chainsaws, keep bees, grow vines, lay hedges.

He is one version, anyway, of the modern countryside: for many years before joining the college he lived off a 6-acre smallholding (net income £10,000 p.a.) raising calves, growing courgettes for the Brighton wholesale market, producing the parsley whose final destination was alongside the sandwiches prepared at Gatwick Airport for first-class passengers. He used to harvest the parsley with an electric hedge-trimmer.

What a wonderful man! The day I spent with him laying a stretch of one of our hedges here was one of the best I have ever had. It was a frozen, exhausting day in the bitterest of east winds, with two coats on, two thick shirts, two T-shirts, hats down over our ears and Boots talking and talking away about the hedge and the plants and the way to do it and not to do it and the way other people did it and the way to put a point on a stake so that it enters the ground and is not split in the process. Over to the east the distant prospect of the fields towards Rye was a stepped succession of bleached-out greys. Above it, the crows from the wood tossed themselves up in the wind.

The hedge we chose was a length of overgrown blackthorn with a couple of hawthorns in it, full of mess and old grasses in the foot, threaded with the ropy cables of honeysuckle and and a big old briar poking up through the middle. 'That'll do,' Boots said, businesslike, with an air of straightforward compe- tence. The whole day flowed like that. It is a gift that good teachers have, to make the arcane seem obvious, the delicate accessible and the skilled no more than a matter of carefully looking and then carefully doing what you have understood needs to be done. Surely the greatest pleasure in life is the process, simply, of getting to know, the sense that your mind, even the whole of your life, is in a small way at that very moment enlarging, like an amoeba putting out a foot and flowing its whole body into and through that extension of itself. Here now with Boots, on the edge of the little field where we had grown our potatoes the previous summer, I felt myself getting to know how to lay a hedge. It was a ratchet clicking up, a stage which, once passed, could never be abandoned. We chose first some thick-stubbed hazels for the stakes and some thin whippy ones for the binders. There was nothing precious about the way he did it. Chainsaw in, sticks on to the field, sawn to a length, tied in a bundle. 'I'm not a traditionalist,' he said. 'I just want to get the thing done, get the light into the hedge, get it growing again at the bottom, make it a living stockproof barrier. If it's easier to use modern tools that's what I'll do and that's what I'll teach you. If it's done right, it'll look right. You want to get it looking right. People always have. In the old days, that was nearly all they had. You could look at what you'd done and say, "Yes, that's well done." It was a way of preserving your dignity when there wasn't very much else that was very dignified in a poor man's life.'

A neglected hedge is a chaos of competing plantlife, a tangle of thorn and deadwood whose energy and focus is at the top end. These are imprisoned trees. Their trunks are stretching out, aiming for the treehood their genes are demanding of them, but leaving the hedge gappy at ground level. A lamb could push between the little trunks and so that genetic destiny is what the hedger has to subvert. A laid hedge is nature slapped back into use.

It is a ruthless business, the precise opposite of the disengaged view that sees a hedged landscape as a comfortable duvet of rural contentment. Laying a hedge is, in vegetable terms, a form of

Part Three

SETTLING

Spring Births, Felled Oaks

AT THE petrol station in the village one day, just as that long first winter felt, perhaps, that it might be on the verge of turning into spring, I happened to look up at a mirror on the wall. It was a moment of self-revelation. A slightly unshaven man in his late 30s was standing beside the pumps, a rather old young man who, in common with many men of his age, was both a little bald and in need of a haircut. The hair that he had was so dirty that it stood up in peaks like whipped egg whites. He imagined, I suppose, that it looked romantically informal, a little wind-swept, perhaps even Byronic. It didn't.

He was wearing a pair of Argyll gumboots, which were muddy around the tops. The trousers of a baggy green corduroy suit were tucked into them. He seemed to be holding them up by putting his hands in his pockets. Under the jacket was what looked like a thick blue workman's shirt. He was putting diesel in a large green Land Rover, which did not, thank goodness, have a bull-bar encrusted with rally-lamps across the radiator but was coated in the splashed mud of which he was obviously proud and had not washed off since he had first bought the thing eight months previously. Pitiable. In the front passenger seat was a nice-looking but rather fat yellow Labrador staring out of the window like a son watching his daddy going off to war.

As a result of the weekly column I had been writing about our life at Perch Hill, a woman from Radio 4's *On Your Farm* programme rang up. I was, she told me, 'part of a general phenomenon'. The words she in fact used were: 'what sounds like the pantomimic quality of life at Perch Hill'. She claimed that our form of rural self-delusion was something that was happening all over the country, 'at least in the pretty parts'.

She threatened to come down and interview us here about our style of farming. I tried to put her off. 'We're not really farmers at all,' I told her, knowing that to be the truth. She took it for the most charming sort of false modesty. 'Oh, come on,' she said, burbling slightly as only producers can. 'Why don't you simply let us come down, have a chat, look around a little and take it from there?' 'I don't think there'll really be enough to talk about for half an hour,' I said desperately, thinking in fact of the dreadful mess everywhere, the chickens in their slum conditions, the ducks in their state of permanent fox-induced anxiety, the ewe that was hobbling about with a bad foot that we couldn't clear up, the chaos of most of the woods, the mud, the mud, the mud. Did I want Radio 4 to see all this? No. It would be like an entire crew of inspecting mothers-in-law coming to stay for a week. Her laugh in response was the nearest to an aural tea-cosy that I have ever heard. 'Oh, really Adam, don't be silly.'

So she was coming. I felt sick and bogus. Sarah was furious. 'I'm not cooking them breakfast,' she said when I gave her the news. 'But they need breakfast for the background sound effects. It's got to sound like a farmhouse kitchen at 6.30 in the morning.' She snorted and left the room. I had the dreadful premonition that when the day came, Sarah would remain sulking in bed while I was interviewed by Oliver Walston, *On Your Farms'* resident interviewer. I'd have to hiss and spit during his questions so that it would sound as if the bacon was cooking in the background. It was going to be hell.

I was in a quandary. What *was* my own view of what we were doing? Part of the time, I knew we were here to recreate a beautiful, traditional landscape, rich with the polycultural detail of orchard, coppice wood, hop garden, pasture and hay meadow that it would have had, say, in the 1870s. And part of the time I realized that was somehow absurd, a meaningless gesture towards a bogus historical accuracy. Why not do what you want to do? Why not make it what you want it to be?

Take, for example, our latest innovation, which had been met with hilarity and disbelief among the neighbouring farmers: sheep bells for the sheep. I found them in a perfectly straight-forward agricultural supplies shop in the suburbs of Palma in Majorca, where they form an everyday part of a sheep farmer's equipment. We got three different sizes for three different notes and could now listen to our small flock as they prepared to lamb at the end of March, their bells rocking gently at the far end of the Target Field. It was beautiful but absurd, Petit Trianon for the 1990s. Carolyn Fieldwick, the shepherd, looked at the sky when I mentioned them. I had yet to tell her that we were also thinking of dyeing the sheep multi-colours so that they would make a broken rainbow across the pastures, pointillist dots on the spread of green. Pretty country, not very 1870s.

There was a paradox I found it difficult to accommodate here. The recreated landscape, the landscape which in some ways seems truest to the place, was in a sense the most bogus of all options, the biggest lie. And the most flippant and superficial of games, the parti-coloured sheep and their bell music from the Mediterranean, were most honestly representative of our own place here now, of our distant, disengaged and in some ways voyeuristic relationship to the land. How honest was I going to be with the people from the radio?

The day came. Oliver Walston, the most famous farmer in England, arrived in his enormous grey Mercedes. The jeans-and-tweed-jacketed, rumbustious, Old Etonian controversialist, who stood with his shoulders back and his chest out like a model of John Bull in a pub, treated Sarah and me gently, even sweetly.

It was captivating. His trick was a sort of faux-aggressive manner which allowed him to get away with murder. Where most people say charming things full of buried hate, he said things that should have been hateful but were overflowing with care and attention. Sarah and I both thought him wonderful.

'What are you?' he asked me over the radio breakfast. The usual packet of Tesco's muesli had been hidden out of sight and plates of agricultural plenty lay there between us. 'A lily-livered, namby-pamby, dilettante aesthete floating about in a violet-tinted world of your own where you want your sheep to be pretty colours and your hedges fluffy? What have I got here, Marie Antoinette?'

'Yes,' I said and went off on to a long blague about the beauty of beauty, how this farm's main crop now was what it looked like, that there was nothing ignoble or contemptible in that, that if this society were not interested in the making or saving of beautiful places, then there was little hope for it. The picture that emerged was of Sarah and me as ignorant amateurs bumbling around 90 acres of the Sussex Weald pontificating about what should and shouldn't be done to the landscape. In other words, a highly accurate portrayal.

Were we consistently inane? Probably. There was a bad moment when I embarked on a lecture discussing the rights and wrongs of nitrogen applied to grassland and all the virtues of no-input management systems. Did I know about soil structure? No. Did I know about the calorie intake required by a grazing cow? No. The biochemical relationship of clover and rye-grass under conditions of climatic stress? No. Nevertheless, I decided to inform a million Radio 4 listeners about those highly fascinating topics. It was the radio equivalent of an undrained bog.

Walston was like the helmsman of an ocean-going yacht watching someone repeatedly capsizing in a dinghy far below. The more I drowned, the more benign his face became. You had to admire the man. After he had gone, the late winter drear seemed even drearier than before: our moment of exposure and then the privacy folding back in.

'It's eight months of winter here,' Will Clark said to me on a dreadful day that February as we stood staring out of a window together at a garden that looked as if it belonged in the outskirts of Chernobyl. 'Yep,' he went on, when he saw from my face that I agreed too much. 'But it won't be long before we're making hay!' He said it smiling, knowing that neither he nor I believed a word of it.

It was March before the sun shone. When it appeared, I felt like shouting hello at it, slapping it on the back and shoving a large glass of sherry into its hand, saying, 'Come on, make your-self at home! Where the hell have you been all this time?' The first days of spring turn one into a brigadier in the East Sussex Yeomanry.

There's a story I always think of in the springtime that comes from one of the deep beech-lined valleys of the Béarnais Pyrenees. A young farmer lived right in the pit of the valley where, all winter long, the mountains above him cast their shadow. It was a place of mist and frost. One autumn he married his sweetheart from another village in another valley and brought her to his cold and shadowed house. They were poor and they struggled through the winter, seeing almost no one and eating no meat. Then, one March day, as he was pulling on his coat in the morning to go out to work, he told her to kill one of their rabbits and cook it, because a good friend was coming to dinner. She duly killed it and cooked it but was surprised to see him coming home at midday alone. 'Where is your friend, then?' she asked rather shortly. He took her by the arm, and showed her the sunlight which at that moment was touching their threshold for the first time since the autumn before.

It is the light that does it, that flood of light, as bleached in reality as the appearance of a midsummer landscape when you've been lying asleep with the sun on your closed lids and you open them to an oddly washed-out world, like a photograph that has been sitting too long on a windowsill. Even in its weakness, spring sunshine is so greedily drunk up. Why is that? I can't quite believe the craving for spring we all feel, so animal an instinct! Nor do I understand how it is that each year the winter seems to grow longer and deeper, the spring more hungered for and, when it comes, richer, more interesting, more of a stimulus, more dominant in the way one feels than it has ever been in your life before. It's as though, as you approach middle age, you become more seasonal, more wafted to and fro on these annual rhythms, less continuous in your life, more susceptible than ever to the conflicting claims of memory and desire. Can that really be the case?

For weeks we had been hunting about, looking for signs of spring. I found a primrose leaf in February the size of a fingernail but crinkled like a Savoy cabbage still half-underground. The grains of soil had folded the tip a little backwards. The hard emergent dagger leaves of the bluebells were pushing through the leaf-litter as if from individual silos. The cow parsley was already there in low, soft-edged pouffes about the size of a dinner plate at the foot of the hedges. Dog's mercury was everywhere in the woods, as well as lords and ladies, and the wild garlic already smelled culinary in the edge of Coombe Wood. One or two of its leaves, bright green, striped dark green, were up and out above the brown wood floor like the blades of soft-bodied assegais.

'That bloody garlic,' Ken Weekes called it. One year, after a winter like this one, when there was no grass left on the fields, the cows had pushed their way out of Target Field and into Coombe Wood, where they saw the alluring bright green of the garlic in the shadows. For half a day the cattle had grazed on the stinking shoots and had then come in to be milked. 'You couldn't even put your face in a churn,' Ken said. 'Phwaw. We had to chuck the lot for three days in a row.'

Apart from that, nothing. There were some leaves out on the honeysuckles and one or two on the elders, but the other trees remained tight and bound in. The hazels and the alders had their catkins dangling in the sunlight and as the breeze blew across them you could see the pollen stream against the light blowing away downwind. But the leaf buds were still hard and inscrutable, genetically wary of late frosts.

Each has its different manner. An oak bud is a heavily armoured thing, protected behind layer on layer of scales like a pangolin's tail. If you flake them off one by one, they come away dry and brown. Only in the very centre do you find the living green, smelling sappy, the minuscule point of protected life. A hornbeam bud surrenders more easily. A couple of flicks at the protective shell, it falls apart and inside you find the cluster of leaves each no more than a 16th of an inch long and covered in silky white hairs like a Labrador's ear. The shape of the future leaf is there. All that is missing is the material. Style precedes substance.

It is the ash, still months away from revealing itself, that is the most defended of all. Its black buds are shielded in points like a deer's hoofs. The outer scales are thick, pointed, firmly anchored and leathery. Pull them away and you find a little capsule of brown fluffy fuzz inside, exactly like rock wool. It is an insulation blanket wrapped around the growing point. Pull that off and you will reach an ash-frond in miniature, the whole frond half the size of a single in-bud hornbeam leaf, still clogged with bits of the rock wool. It is tentacled like a sea-anemone and looks as if it should belong on a coral reef. Such care, such details! Perhaps amazement isn't really enough of a reaction. But if not sufficient, it is at least necessary. That is what spring-time is: gratitude married to amazement.

A double crisis, long predicted, and even longed for, started to close around our lives. That March, Sarah and the sheep were all, in a miraculous piece of synchronicity, on the point of giving birth at home. I was waking up with my teeth clenched. I was yo-yoing between cow shed and sitting room, sitting room and cow shed, in a stew of vastly enlarged paternal concern. Twenty mothers in my care! Sarah and I had erected a 6-foot-wide, 4-foot-deep swimming pool in front of the fire in the sitting room. All furniture had been pushed to the walls as though for a dance. By 1 March, Sarah had already had two full-blown 'It's coming' crises and those were somehow worse than the real thing. I felt in those Phoney War days as though I were a Battle of Britain pilot sitting on an armchair arranged next to the runway, my Mae West around my neck, nonchalantly smoking, while my insides were doing the can-can. Rosie, our two-year-old daughter, woke me up one morning to ask if I minded if she cut off my head with a carving knife. I said that would be fine.

Then there were the sheep. They were due to lamb in a couple of weeks but some were bound to be early. Luckily they were unable to say when they were having an 'I think it's coming' crisis. Or if that was what they meant by their bleating and shuffling at six in the morning, I just ignored it, gave them some more hay and told them to shut up.

It was vital that no hint of pregnant sheep came anywhere near pregnant wife for fear of disease spreading from one to the other. I have never washed so much in my life. Sitting in the kitchen, I had lessons from Carolyn Fieldwick, the shepherdess in boiler suit and woolly socks, telling me in precise and careful detail what I had to do. There would be the three-hourly, twenty-four-hour-a-day inspections of the ewes from the beginning of March until mid-April. If I found one whose womb was prolapsing ('You'll see a very red, pinkish blob the size of a fist coming out'), I had to turn the ewe over, wind baler twine round her middle, 'push everything back in' and then tie it there with a special bit of kit I had to buy.

What if the lambs were coming out head but not feet first? Reach in to get the feet out but check that the feet belong to the lamb whose head you can see. Twins get muddled up together. Didn't Ted Hughes write a poem about pulling on a lamb so hard that its head came off in the womb? What if it was coming out backwards? What if the second lamb was all muddled up with the 'bag'? What do you do about the navel? What about triplets? How do you get triplets all to suckle? What if the mother dies? I was in a state of tense, exhausted paralysis.

The weather made it worse. March, it has to be said, is the most vindictive month. There is a catty, cold-blooded compassionlessness about the way it promises you everything and never delivers. March, in fact, is a liar. It lets you pretend for a while that England is a northern limb of that benign southern Europe where apricots coat the walls and life is lush and generous. But March comes back, old and deceiving, turns the heat and benignity off, replacing Provence with Spitzbergen, and prunes away at the loose-limbed hopes the warmth had engendered. Spring had come, with its usual severity.

Three in the morning, two weeks later. I'm in the cow shed to see that the sheep are all right. The night in as dark as it will ever be. The south-east wind is cruising in like a shark off the English Channel, ten miles away. It's coming in through the spaces at the top of the barn doors and out the far side. The water is frozen in the buckets.

The ewes are in here, nineteen of them, as pregnant as a fleet of East Indiamen, laden to the gunwales, bulging with themselves. Some of them look as though they had a pair of saddlebags strapped around their middles and when they lie down, as they are now, their vast, filled midriffs pool out on either side in an ocean of motherhood and fecundity.

Their time is due. Roger, our Suffolk ram, now grazing with two young ram lambs in the field on the far side of the road, did well in the autumn. Only one old black ewe, well past her prime and possibly barren, is not in lamb. She's in the bull pen now and she looks out past the hurdle at the door with an air of abandonment and age. Her black wool is grizzled; she won't last the spring.

That's not what it's like in here. Despite the cold, despite their laden condition, the ewes are lying out across their thick bed of straw in pure horizontal contentment. We've fed them well, for weeks now, on quantities of ewe nuts. Half a ton has disappeared down their gullets, not to speak of 2 acres-worth of the hay we made last July in Beech Meadow, good 'blue' hay, meaning there is still a certain greenness to it even at the tail end of the winter. The sheep have had nothing but the best and they look marvellous on it. There is something about them which reminds me of a plateful of gnocchi, a rounded warmness, comfort made flesh. Their chins are lifted in the attitude of sheep at ease and a low snorting sort of snoring is coming from their nostrils. The expectant mothers are happy.

One has already done what she needs to do and is over the other side in an individual pen with her lamb. It was Sunday lunchtime. She had been shuffling about all morning, looking, as Peter Clark so precisely described it, 'a little sheepish', and then during lunch must have delivered.

We found her with the lamb still smeary with the membranes at her feet and the afterbirth still hanging from her. It was all so normal, so griefless, so prosaic in its way, so without agony that it now seems absurd that I should have gone in for so much apprehension. This was as it should be: ewes in good condition deliver easily and have the appearance afterwards of nothing having happened.

I picked up the lamb to put her in the pen. The little thing felt just as if someone had broken eggs all over the wool. The ewe followed us. They licked and nosed each other. A lamb is a survival machine: a big head, a big mouth and four stocky black legs out of proportion to the sack of a body which joins these standing and eating parts together. The ewe had a full udder, and the lamb soon found its way to suck, wriggling its tail, the instinctive drive at work, the vital colostrum running into the gut. Survival.

A sheep has no face, no screen on which its mental state can be read as ours can in such detail and with such immediacy. You look at a sheep and see a certain blankness: no pleasure, no pain, no grief, no anger, no delight, no regret. But if there's no face, you can at least read its body and, unlike the sisters still waiting for their birthing moment so relaxedly in their communal pre-crèche, the mother with its lamb was obviously in a state of acute anxiety. For twenty-four hours after the birth, whenever I came in to see how the lamb was doing – my own anxiety, needing this thing to survive, not to die on me, not this first one – I found the mother standing alert, eyes big, defensive, stamping her front feet as I approached the pen or picked up the lamb to look at the navel and the shrivelling cord or to feel its, gratifyingly, filling belly. The ewe is tensed to protect her own. She is a servant of her genetic destiny. Her life can only be dedicated to these fragile, transitional moments on which so much hinges. So this instant, in the pen with the hours-old lamb, with the tautened presence of the protective mother, this is one of those moments when you come close to 'the blood of the world', to the essential juices running under the everyday surface of things, when the curtain is drawn back and you find yourself face to face with how things are.

It snowed the following night and in the morning the east wind blew thick drifts of it off the field and into the trench of Willingford Lane, blocking it. Sarah was due to give birth at home any day and in these conditions no doctor or midwife could reach her. I drove up the lane in the Land Rover with a spade and shovel and for four hours dug a way through the snow, cutting the route which an emergency would make necessary. It was the opposite of digging a grave, cutting a life-path through the snow, a car wide, three or four feet deep for about 50 yards along the lane, just on the crest where it was open to the wind and where the snow had swelled into bulbous goitres and growths between the hedges.

The next day Sarah gave birth to our second daughter. The contractions began the evening before. I half-slept, waking, stoking the fire, sleeping again, refilling the pool with warm water, and Sarah bathed there throughout the night, calm and easy. Sarah's sister Jane was staying and while I slept, she looked after Sarah. An old friend, Patricia Howie, was here too, to look after us all. The midwife came at about six. By nine in the morning the whole process had steepened and deepened. In quite a sudden way, with the growing contractions, it reached a huge and passionate intensity. Sarah looked in this extremis like one of the sibyls on the Sistine Chapel ceiling, a vast being in pain. I could hardly recognize her. I was amazed by it. The sheer hurt of the delivery seemed at times to balloon out from her to fill the whole room, the whole house, the whole of here. She was shouting louder than I had heard anyone shout before. Human birth, when seen at home, when suffered by someone you know and love, when not dulled or interfered with by the dislocations of hospital and its comprehensive anaesthesia, is a vast and violent thing. A husband, an observer, can do little but stand and watch, gormless in his irrelevance, awed by the sort of instinctive courage this moment summons, bewildered by the sheer scale of an experience which little else in life can match and finally swept away and dissolved in the relief of its ending, its happy ending in a daughter who was well and who would survive. Sarah and Molly were well. But why should it be like this? Why should human birth take such a toll? Why can't a woman give birth like a ewe? At just after ten in the morning, Molly was born into the water of the pool, scooped up and out on to Sarah's breast and I wept with the relief of it. An hour later they were both in our bed and I put flowers all round them and tall branches of hazel cut from the wood with the tops of the hazel fronds bent down by our bedroom ceiling, so that all round the bed the catkins hung down over them both, a bower for my family.

This of course was how it should be, an unbroken transition from womb to life, and as I looked at Molly that morning, still blinking in the shock of her extra-uterine existence, I realized that in some ways she was still being born.

Only one Molly, but endless lambs. By the time she was a week old, 10 of the ewes had given birth and they had delivered 18 lambs. Most had been twins, but there had been a set of triplets and one or two singletons. Only one lamb, one of a pair of black twins, had died, a week after Molly's birthday, at breakfast. It had been born the afternoon before and looked all right to start with, if very small, and I didn't notice anything the matter when I checked at about midnight and again at five the next morning. But at eight o'clock I found the poor black thing suddenly crashed out and hopeless, lying all wrong on the straw, its body too heavy for itself. When I picked it up, the head hung down at the end of a muscleless neck and its body slumped in my arms like a little pietà.

We brought it into the kitchen to warm it up and bottle-feed it. The milk went in and it seemed to be swallowing but that can only have been an involuntary impulse; the animal was already virtually dead. It had probably gone too far by the time I found it, for some reason neglected by its mother overnight, even though it had been feeding well enough in the evening. Its heart was still fluttering when I first picked it up in the morning but within half an hour the pulse had gone and its whole body had moved across the unnoticed line between 'ill' and 'dead'. We buried it near a ewe that had died the previous autumn. The dribbled bottle-milk was still coating its chin in a veined white slick as the lumps of clay fell and bounced on the body.

Death at lambing is only to be expected. In fact, we got away lightly. The ewes were all fine and the rest of the lambs well and lusty. The Fieldwicks, with 400 ewes, which were not meant to have started lambing until April, had already lost four of them. They had been found dead in the field. The year before, for no reason they could tell, they had 60 barren out of the 400 and trailerloads of dead lambs carted away to the tip. This for them was the anxious time. The week before, a ewe had been delivered of triplets prematurely. One of them had died almost straight away. The second needed bottle-feeding, the third was sucking well from its mother. Two nights later Carolyn found the ewe herself dead in the field. She had rolled over and crushed the one lamb that was making a go of it. The prospect of exhaustion and failure hung around the whole business. I could only thank God that this was not the way I earned my living. I knew young farmers around here, struggling in the grip of this, man and wife working all hours, crucified on the cost of the grass keep, with a quota only for a small proportion of their flock, their young teenage children dragooned into helping when they would otherwise be at school, the strain telling on everyone's face. I was always amazed at how young these old-looking people were. People I thought of as much older than me turned out to be far younger. I felt coddled by comparison, padded by the soft-ness of a sort of life which their whole life's work might in the end bring them. And only then if it went well, if they stuck at it, if disaster did not pick away at their sliver-thin margins.

The survival rate of lambs, the cost of rations for the pregnant mothers, the condition of ewes after lambing: these things are the determining factors in what their lives would be like this year, next year and for ever. To be so dependent on the uterine workings of another species! I said to one of them one day, his face taut with exhaustion, what hell the life of a small, under-capitalized sheep farmer must be. It was a mistake. He sat up, flicked his head half sideways in the way a cockerel might, and said, 'Why? What's wrong with it? I like my life. I like it a damn sight more than I'd like yours.' We are all tender.

For us, though, it was the sweetest of times. Rosie played

among the lambs. They danced and pranced together in the orchard outside the house. Molly peeked like a mouse from her swaddling. Tom, William and Ben cradled their sister in turn and the lemon-yellow sun shone on our lives.

A few weeks before Molly and all those first lambs were coming to life, I had forty oak trees cut down. We wanted them for a building, to restore the oast, the other side of the yard from the house. Forty oak trees! I could imagine them growing in the sort of open circular grove the Greeks would have admired, gradually filling out over the course of this century, swept by the wind, a grassy lawn beneath them. And I had them cut down.

In fact, I never saw them standing. They had been growing in a wood at Ashburnham, a few miles south of here, and I could imagine the rawness their removal left, a stretch of land looking as a gum feels after a tooth has been drawn: the awful, shocked absence at the site, a ragged-edged nothing where something should be, a place whose gruesome softnesses your tongue can only tentatively explore.

I first saw our trees in the timber yard belonging to Zak Soudain a few miles away from here near Broad Oak. The wood was still wet and very green, as the word is, although the real colour of sodden, recently felled oak is an orangey yellow that verges on pink. Where a chisel hacks at it, flaking the fibres of the wood, the nearest thing to green oak that I've ever seen is the raw flesh of a salmon. The lichen was still growing on the bark of the enormous, horizontal trees as they lay in Zak's yard, so you could still tell which had been the sunny southern side and which the northern when they had been standing in the wood. Despite that, though, the oaks had already moved over from one side of the equation – the handsome shape of the living tree – to the other, a dignified after-life as timber.

I didn't feel a trace of regret about having the trees felled at the time, and still don't, only excitement at what was to become of them. The new building was to have a green oak frame and be clothed in oak weatherboards. The structural techniques, the joints used and the material of which the building was made were all identical to those used to put up the farmhouse 400 years ago, the hay barn about 180 years ago and the original oast-house about 130 years ago. Every one of these buildings has used the local oaks for their main structural members. Our new building would be the fourth time in four centuries that someone at Perch Hill Farm would have had a fair stand of oak trees felled and put to use. And each time the woodland was not lost but cropped.

There is rationale to those time intervals. They are governed by the market. Each of the moments that one of these buildings was put up marks a period of optimism and expansion in the farms of the Sussex Weald. In the sixteenth century, as prices rose under population pressure, it became profitable to farm even the more marginal lands like our wet, woody clays: time for a new farmhouse at Perch Hill. In the Napoleonic Wars, blockades created shortages and shortages created cash for suppliers: time for a new barn. In the 1860s, booming populations demanded oceans of beer, railways allowed national distribution and hops became the new cash crop: time for an oast-house at Perch Hill.

And now? The market is no longer in such obvious commodities. Hops, corn and milk can now only be produced commercially on a scale to which this landscape has been unable to adapt. Ten years ago there were five active dairy farms on our lane. Now there is none. Every farmhouse is occupied by one urban professional or another. The farming is done by people renting the grazing from elsewhere. The City, land agency, the media, old folks' homes, fashion, the law: that is where the money is now and that is what now owns Willingford Lane. The only crop this landscape can viably produce is beauty and the only thing it can sell is itself. It is doing that very well. New money from the cities has arrived and once again, in a green-oak building, Perch Hill is getting the reward it deserves. Ever since this farm was first cut from the forest, the market for its produce has been urban. There is no radical break with the past in what we are doing. This is the Perch Hill way: an influx of urban money means green-oak buildings, built to last for centuries. The 1590s, the 1810s, the 1860s, the 1990s: these are the steps in the graph, the moments Perch Hill takes another step forward.

To begin with – doesn't everybody begin the story of a building project with that phrase? – all was marvellous. The trees were cut into the shapes required for the giant frame. Others were sliced by a cheese-paring saw into the long, feather-edged weatherboards that were to clad the walls. The money we had would be enough. We saw the building changing day by day, the holes for windows opened, the brickwork growing for the new upper storey to the roundel. But then, creepingly, apparently unawares, delays began to appear. People wouldn't turn up. Alterations turned out to cost much more than expected. An air of catastrophe hung above the scheme. Its noise and disturbance began to eat at our sense of well-being.

One day that summer, a minibus full of semi-antique ladies from Bexhill, most of them wearing the kind of maroon felt hats that look as if they should be the central structural element in one of Delia's Light Afternoon Sponges, pulled up outside our farm gate. The bus had parked just opposite the chaos of our building site. The job was now many, many weeks late. Mess lay everywhere. No one was at work.

The tour leader on the bus, microphone in hand, pointed out of the window at our building and said, 'There you have one of the old oast-houses of Sussex and Kent in which, in the old days, they used to dry the old hops. Lovely things; you could always smell the drying hops for miles away downwind.' The bus aahed. 'But many farmers,' he went on, 'are now finding oast-houses rather inconvenient and, as you can see here, are dismantling them to make way for more suitable buildings. It's a shame but many farmers are struggling to make a living in this part of the country and after all it is a free world.'

I stood there in my gumboots, listening to this open-mouthed. The Fruit Compôtes in the bus in front of me all turned their attention from the building – which was costing as much as a Ferrari Testarossa to put up, entirely funded by some particularly acute investments I had made in the 1980s, the farsightedness of Barclays Bank, Hammersmith and the blessed generosity of my own father – and, in a single, coordinated gesture of patronizing benevolence, directed fourteen pairs of twinkling glacé cherry eyes on me. I had to turn away.

What was done of the building was, on the whole, beautifully done. If you could ignore the fact that, six months after it was due to be complete, it was still unfinished, that the pointing of the new brickwork seemed to have been done by someone who had only ever previously worked with play-dough, that there were no doors, that the windows had no catches, that there was no floor upstairs, that four of the seven lights they installed one week were not working the next and that my brother-in-law thought the whole pitch of the roof was wrong, it was really very good indeed.

There was the slight problem that the two men who were meant to be the main contractors on the job, and whom we engaged only because they were so attractive (one in a rather saturnine, agonized, Übermensch-under-strain way, one a fresh-faced male version of the freckled English rose), had fallen out with each other so badly that they were on the brink of a vicious legalo-financio-emotional-hurt-and-betrayal dispute which might or might not end up in court. The saturnine one of the pair, who had a Heathcliff-goes-clubbing look to him, and wore a sort of silver anorak that seemed to have been cut from the fuselage of a 1952 USAF strato-cruiser, had arrived on site with a new black eye on two different eyes in two consecutive weeks and had said both times that he had walked into his car door the previous evening.

All that aside, the job went rather well. The blips and hiccups, the overruns and punch-ups, the walkings-off the job, the mysterious disappearance of a septic tank one night and the discovery that the oak floorboards we had already paid for were so wet that if they had been nailed down in that condition they would all have buckled into a model of the North Dakota badlands within a couple of months, all that was nothing more than what you might expect. That was what life should be like – a little spurty, free enough to go wrong.

No one else could understand this point of view, particularly Sarah and the bank manager. Mainly to satisfy them, I did, on a couple of occasions, lose my temper with Heathcliff-in-clubland. It was cynically done. I was at the end of my tether anyway and it seemed to me that screaming at the poor man down the phone was better than kicking the dogs/cats/sheep/ducks/chickens/walls/children and would go down well with the wife. It didn't have the right effect at all. Heathcliff came round, explained the problem away perfectly and looked hunkier than ever. Sarah ended up admiring him more than me and I thought for an alienated minute they were going to go out together that evening to a fantastic place he knew in Croydon – 'brilliant, it's a warehouse which has been lined inside with the façade of a Renaissance château'. The job continued to progress at an inch a week.

There was one little thing about the building which remained a niggle, which ran against the lovely freedom-is-beauty gospel to which the whole of the farm had become dedicated. The new building had two bathrooms in it, one upstairs and one down. Both had large windows opening on to a view of a grassy bank and, over to one side, the chicken slum where the three survivors out of our original flock of twenty hens and one cockerel were scratching out their tragic lives. Both these windows opened fully. This is one of the breeziest places in Sussex and there is never any shortage of fresh air. If you had your priorities upside down, and if things ever turned financially disastrous, this would be a prime spot for a wind farm.

Despite the natural gush of air past our fully opening windows, despite the fact that we would probably like to open those windows to enjoy the sort of air that we had come here for, we were obliged to install in each bathroom an electric extractor fan. They are ugly little things, plastic, louvred squares which turn on when the light turns on. I hated everything about these fans: the look of them, their noise, their enforced presence, their waste of money – £40 each – their attitude, above all, of tidying up our lives for us. I could imagine lying in the bath in the future looking up at the fan whirring its little whir above me and thinking, 'Go away, I hate you.'

It was the 1991 Building Regulations (1995 Edition) Part F (i) which required me to install these little things. It was all to do with 'interstitial condensation' – or damp in the rafters. I couldn't, first of all, be trusted to open the windows myself, which was irritating in itself. But from talking to the Rother District Council Building Control Officer, it became clear that this little plastic imposition was symptomatic of a much larger phenomenon. Houses used to breathe. Surfaces and materials were in some ways permeable to the wet; there was more of a flow between inside and outside. Then came central heating, then the require-ment to preserve more of the heat so expensively created, then thick insulating materials, then what the RDCBCO called 'the house like a kettle', all the hot wetness from kitchens and bathrooms held within this sealed container. Then came the requirement for the electric fans because people could no longer be trusted to open their windows and break the precious seal.

It was the classic example of the way in which people had removed themselves from their natural environment. Every step follows logically from the one before until you suddenly look round and find you are halfway up a cliff and don't like the feeling at all.

It is not that it is ugly; a light switch or a plug is ugly and I don't mind them. The fan was horrible because it was the mark of alienation, of a sterilized, cut-off tightness, of a ludicrously unnatural way of regulating your life, of over-prescription, of denying yourself the feeling, that wonderful summer feeling, of the breeze against the skin, which is one of the reasons you are alive in the first place. So once the fans were in, I took them out and I felt the building, the beautiful, creaking, appallingly expensive, debt-creating, naturally sweet-smelling, oaky heaven of a building, sigh with relief.

It is in that fan-free place that I have written this book. I live with the wood day by day. The timber dries and as it dries it shifts. As it shifts it splits and as it splits it creaks, as though the whole thing were springing apart. From time to time, unexpectedly, at a quiet moment, the whole frame creaks, not in an old or easy way but with a sudden, high-pitched jerking under stress, a shriek of wood, a spasmodic movement, in the way that earthquakes happen. Over many months or even years, the tension builds and then, bang, catastrophe theory at work, it becomes too much. The pieces move not with an easy, oiled constancy, but in a little convulsion, a twitch.

That moment sounds not wooden at all, but polystyrene. You might hear in that agonized squeal the sound of all the torture that preceded it, a final, desperate outburst of the wood under strain, its elasticity stretched quite literally to breaking. Every time it happens, I look up at the jowl-posts and tie-beams, at the scissor brace in the apex of each truss, and see nothing. The building, an invisibly clenched and tense thing, where the fibres in the timbers are tightening and stretching against the pegs that hold them together, remains inscrutable. It looks stiff, solid, immobile, as reliable as buildings are meant to be. 'What, me?' those impassive beams ask, as I interrogate them about the noise they've just made. 'Can't you see, we are what we always were?'

In Deepest Arcadia

As SPRING thickened into summer, and both the hay and the corn started to coat the country in a deep green pelt, a kind of sexiness began to seep out into the fields and their hidden corners. The height of early summer turned into the most lustful moment of the year. Driving down Willingford Lane in those lush green weeks, dropping from Burwash Weald to the bridge over the Dudwell and then up through the outer patches of Dallington Forest, moving from sun to shade and back again, past our farm and on towards Brightling, you would find cars parked in the evening in the tucked-in gateways, reversed half out of sight among the cow parsley, an air of privacy and closure about them.

They were always young men's cars, the sets of wheels that could be afforded rather than desired: a rusting Escort estate, a brown Capri, and never anyone visible in them. Each one was a strange prefiguring of the way those old men's cars, the brown Granada, the 'Autumn Gold' Austin Vanden Plas with walnut trim, were always parked on similarly beautiful evenings, not in the quiet corners but at the viewpoints, on the Downs and the higher places in the Weald, the bonnets aimed at the landscape, the old man and his wife sitting calmly in the two front seats watching Sussex as though it were an intermission in Thursday night TV. They pass the thermos, the wife worries about the children, the dog farts silently on the back seat and the husband thinks his lustful thoughts about his youth and all those never-confessed-to lovers. 'Men are April when they woo, December when they wed: maids are May when they are maids, but the sky changes when they are wives.' *As You Like It*, the route-map to the pastoral idea.

June is the month for outdoor lust, now as it always has been. The famous song sung by the two pages in *As You Like It* is a precise description of the sexual habits of the rural working class in early modern England:

> It was a lover and his lass,
> With a hey and a ho, and a hey nonino,
> That o'er the green cornfield did pass . . .
> Between the acres of the rye,
> With a hey and a ho, and a hey nonino,
> These pretty country folks would lie.

Everything about this is accurate: they make their way right to the other side of the growing cornfield, away from the invigilating police state of the 16th-century village, and there find the privacy they crave. You hardly ever see rye growing in England now, but it is the lovers' crop *par excellence*, six feet tall by the middle of June, a wall of protective green. 'With a hey and a ho, and a hey nonino' sounds innocent but it isn't. 'Nonino' is 16th-century for 'a bit of the other' and even 'hey' has a lustful tinge to it. In Shakespeare, the word 'country' is always enriched by the pun it contains. The whole movement of the song, through the fluffy acres of the fields and on into the safe and private lying place, is sexual. It is a hymn to sex as summer heaven.

It is easy to forget how thick with all this the landscape still is. I was looking that year into the history of the woods that surround our farm, the woods that formed the rich and numinous background to all Kipling's *Puck of Pook's Hill* stories. That was my idea of them, the source of a psychic magic in which Kipling revelled, until I began to ask in the village about what the woods had been like before the war. The old men I spoke to would begin politely enough, mention their work in the wood, the trees that had once grown there, but their eyes would light up, narrowing and brightening at the same time, when they talked about 'going courting' in High Wood: no better place in the summer than High Wood or Leggett's Wood, the mossy banks in the old sunk roads, the open ground beneath the beeches where you could spread your coat on a bed of bracken, the beech mast being too crinkly and spiky if it's under a person's back, and you don't want them to feel uncomfortable, do you, you don't want them distracted by the prickles . . .

Those beech trees remain the poignant memorials, standing huge and isolated among the twiggy birches. One of the beech trees, in particular, is an incredible being, a balloon, in the newness of summer, of fresh lime-juice-green leaves, with two tennis courts of shade beneath its branches. The muscled limbs are clothed in elephant hide and the dress of windblown leaves acts the feminine to that massive masculinity. Of course it is the place for seduction.

The trunk is carved with the initials of forgotten lovers. The bark is cracked and pitted inside the rough-cut serifs and the places where an O or a C have thickened with time. Here and there, moss inhabits the carved-in words. Hearts enclosing girls' initials have been stretched so wide by the swelling of the tree that they look like one of those gurning grimaces made by five-year-olds, fingers hooked in the corners of the mouth and dragged out sideways across the width of the face. Inside that gruesome cartouche, the initials are now illegible, pulled beyond understanding, no more than a smeared-out mark which the tree has done its best to erase.

It all brought back memories of outdoor love affairs decades ago, that odd sensation of the breeze in unexpected places, the disaster with my first-ever girlfriend, on a hillside in northern

Spain. She was an Argentine; I hardly knew her. She had no English to speak of and so we scarcely spoke except in inadequate French. 'Don't hurt me,' she said that afternoon as I was looking over at the view and I thought she meant I must not treat her badly, as all the signs were surely pointing in that way. 'I won't,' I said solicitously. 'No, don't hurt me,' she said with a little more emphasis and I equally fiercely said I wouldn't, of course. 'No, NOW,' she screamed in my ear and pushed me away. It turned out that the way I was leaning on her was pressing her far shoulder into a small, invisible but obviously rather prickly thistle.

At Perch Hill, the thistles were well into their wild annual career and I didn't like it. From my desk, I could look out across the rising bulk of the Cottage Field towards Coombe Wood. The field looked wonderful, a perfect sward, next winter's hay in the making, as invitingly edible as a plate of rocket and watercress salad. But the appearance was a lie; its reality was a nightmare of weeds. Walk across it and the luxurious softness disappeared. Your boots crunched at each step as if on shingle but what you were treading were thistles, many thousands of them, still little more than horizontal rosettes at this stage, nestling invisibly in the grass but soon to start their growth upwards. Where there were no thistles there were docks, and where there were no docks there were nettles; where there were no nettles there were brambles, and where there were no brambles, there were dandelions.

Before we came here, I had a supremely haughty attitude to grassland. If, walking around England, I came across fields like ours, I would have one of two remarks to pass. It was usually: 'Poor management, *very* poor management. They don't really know what they are doing.' The fact that I didn't know what they were doing, or what they were meant to be doing, or what I would have done in the circumstances, or what sort of management history would lead to this sort of weed problem, did not stop me from passing judgement. I now realized why farmers hated people like me.

At other times I would say, 'Of course a weed is just a frame of mind.' That saying exists in the same sort of sententious mental lay-by as those notices at the entrances to US National Parks: 'Take nothing but memories, leave nothing but silence, speak nothing but peace.' I once picked up a sandstone pebble in the titanic desert emptinesses of the Utah Canyonlands – it was beautiful, with the ripples of a red Jurassic beach on its surface – only to have it confiscated by a National Park Service Ranger, a woman with brown curly hair and a revolver, on the basis that I was 'disinheriting the generations that come after'. Forget *objets trouvés*; they have slipped beyond the bounds of the acceptable.

There is an idea that at one time, when the people of this country were still at home with the ways of nature, the plants we now see as weeds, of which we know nothing, were seen in their true light, useful as food or medicine. Nettles cured stomach upsets and made excellent cloth; the fruit of the bramble has been found in the stomachs of Stone Age men preserved in Irish bogs; young thistle stems, blanched and peeled, were eaten like the heart of an artichoke and were said in the 16th century to be 'sovereign for melancholy'.

That is the side you always hear about nowadays, yet another measure of our fall from grace. But it is no more than half of it. Anyone who has read that wonderfully encyclopaedic treasure-house *The Englishman's Flora* by Geoffrey Grigson, a work of love and scholarship from which everyone else has always cribbed whatever knowledge they pretend to have, will get the fuller picture. It provides a strange enlightenment.

Every year parts of the Middle Shaw are dominated by the dreary, tiny-flowered, big-leaved plant called dog's mercury. It looks boring, it's not useful, it's a weed and it seems to outcompete bluebells.

But what about that elegant, alluring name? Until I read Grigson I always thought I must be missing something about this plant. Not at all. This woodland mercury is a perennial, highly poisonous, both emetic and purgative, one of the lowest of the low, good only for dogs. Dog's mercury means 'rubbish mercury', 'weed mercury'. What a liberating recognition that is! The name reveals that pre-modern people hated weeds too. Grigson lists 70-odd names of wildflowers which have this dismissive 'dog' element in them: dog jobs, dog-cock, dog's mouth, dog stalk and so on. As for cow parsley, which was then appearing on the banks of the lane and which everyone loves, that too is a historically despised thing, its tauntingly full but useless growth associated with the devil. It's the devil's parsley in Cheshire, dog parsley in Hertfordshire, gipsy's parsley in Somerset, hare's parsley in Wiltshire.

From admiration to contempt, from exploitation to at least local extermination, the historical attitude to wild plants covered the full range. Po-faced piety about the natural order didn't get a look in. So, I said to Ken Weekes one morning, what shall I do about the thistles? 'Hammer them,' he said. 'Hammer them as hard as you like.' So that's exactly what we did. Will Clark drove the tractor, I bought a giant flail topper with a 9-foot cut and Will, in sweep after sweep, beheaded every one in every field. He let me have a go, standing on the side of the field with his hands on his hips and shouting up as I passed about the throttle or the straightness of the line which the topper was leaving. The satisfaction of this: looking back, one hand on the wheel, one on the back of the tractor seat, a swept wake of grassland emerging behind you, the blades of grass laid low by the topper as it passed and now sheeny in the sunlight, orderli-ness, a place which has been worked.

One morning that summer a man knocked on the back door. He wore a sort of yellowish canvas coat with a corduroy collar and took his muddy shoes off with deliberation before coming to sit down in the kitchen. He had something of an ex-naval air: affable, polite, attentive.

He too, he told us, looking at the half-wreck of the oast-house outside, had gone in for building works. Oh, the headaches! He had discovered terrible subsidence and had been forced to pour money into a hole in the ground, far more money than he had. That was the reason he was now working for Orange, for Hutchison Telecom. Eyebrows up. His job was to find sites for the masts that would give the Orange network the coverage it had to provide. Would we be interested?

'Tell me about it,' I said, thinking, 'No, not here, never.' He showed us a series of photos of the masts, 50 feet high, surrounded by things that look like giant chest freezers around the base, enclosed within a chain-link anti-vandal fence and surmounted by the aerials: big, dominant and ugly.

The poor man made no attempt to pretend they were anything other than dreadful. His word, in fact, was 'beastly'. He was charming, disarming even. I wondered, but didn't ask, if this was the trained technique: spit the worst out early on, show them you understand what they are frightened of and then offer the blandishments. So what would the deal be?

'We'd make an agreement for ten years,' he said. 'We'd rent a patch of ground 10 metres by 10 metres from you, and we usually offer £1,250 a year for that.' My face looked like a slot machine as the dollar signs rolled. £12,500 for a patch of Beech Meadow 30 feet square? Oh yes.

'We had the Mercury man round here last year,' I said, lying, repeating something someone else had said to me about a visit to their farm months ago. 'He offered £2,000 a year.' 'I thought that might be the case,' our Orange rep said, 'and obviously, as an annual tariff, we'd have to match that. Yup. And we could say that it should rise with inflation.' Already £20,000. And he might go up a little more if I dug my heels in. But I was mucking him about. I never had the slightest intention of having such a monstrosity looming over our fields. It would be like putting callipers on the leg of a child.

'Now this is the bit I don't like,' the rep said, smiling like an old friend. 'It always sounds threatening but it's not meant to be. If you don't agree to have the base station on your land, I'll have to go to your neighbours and see what they make of it. And obviously, if they think it's a good idea, I'll be going ahead with it with them. You do understand, don't you?'

Oh Jesus. There is a thin sliver of land that runs through the middle of the farm which we don't own. Its owner lives miles away. The prospect lurched up of a 50-foot hideosity on the bank above the farm and no compensation.

We both allowed the talk to burble on around this aching pothole. He mentioned 'statutory obligation to establish a national network'. I mentioned that this was statutorily an Area of Outstanding Natural Beauty. He knew that and said about planning applications being granted on appeal by the Secretary of State. I said that if we didn't want it, we'd do everything in our power to stop it happening, and I meant everything. He said he understood and I said, smiling, shaking hands, opening the door, that I'd be in touch.

I wasn't. He rang, but I wasn't in. He wrote, offering 'to sugar the pill with some modest improvement on the annual fee of £2,000 I suggested and possibly throw in an Orange, with a year's basic tariff (15 free minutes a month). At least it should work well!' I didn't reply but I guessed that his other options were somehow closing and that waiting was working. Then, months later, the letter we wanted. The radio engineers, with their 'computer modelling tools' had decided that somewhere 'further west' would better suit their purposes. Cheers at breakfast. The future's bright? The future's Orange? Not here it wasn't.

While dreading the irruption of Orange into our lives, seeing wherever I went the spectacular unregulated ugliness of the mobile phone towers (and enjoying for the first time all the pleasure of having a mobile phone myself), I had been looking forward to something else, not the disruption of the local by the national but in many ways its opposite: the 49th Annual Heathfield and District Agricultural Show. I had been asked to be one of the judges. The invitation had arrived months before from the charming Mrs Berger, Hon. Trade Stand Secretary, and as soon as I opened the envelope, I knew that, as far as the Weald of Sussex was concerned, I had arrived. A Heathfield Show Judge!

For weeks, I was modest about it at parties but privately triumphant. The badge, a hexagon of stiff, burgundy-coloured cardboard with the word 'judge' stamped on it in gold, came through the post and I tried it out in front of the long bath-room mirror with a variety of different suits and ties. Dark blue was obviously wrong. Hairy Harris tweed was absurd for early summer. It could only be the cowpat-green corduroy. Come the day, I immediately realized that my fellow judges were more impressive – I want to say realistic-looking – than me. John Bines Esq. was once a government expert on the feeding of dairy cattle and then Chief Executive and Secretary of Newbury and District Agricultural Society. Mrs Valerie Chidson was Chairman of the Wealden District Council. She had, of course, unrivalled local knowledge and wore a great badge around her neck like a mayor, with a long flamboyant apricot and orange silk scarf floating above it. I was described in the programme as 'A. Nicolson Esq., Journalist' which looked disreputable. Why not 'landowner, landscape theorist and visionary'? That word 'journalist', it's no good. I was told by an insurance agent once that I should never, ever mention what I do. Only 'fairground booth operator', he told me, was consid-ered a more dangerous risk.

Anyway, Mr Bines, Mrs Chidson and I, in a couple of hours, were walked around the two hundred-odd stands of the show.

They were laid out in broad, muddy streets across a hillside outside Heathfield. The setting-up day had been rainy and the trucks had turned the field into a quagmire. But now the sun was shining on the stands and tents; on the enormously fat horses being trotted up and down by enormously fat men in suits; on the Side Saddle Concours d'Elégance where double chins wobbled beneath antique veils; and on the Heavyweight Hunter class, capable of carrying fourteen stone and over, which Miss S. Waddilove had come down from Newmarket to judge, along with the other Ridden Hunters.

In the cattle rings, the junior handlers, the boys and girls, were struggling with their recalcitrant calves. 'Will you bloody well come on!' one tiny boy in his pristine white coat said out of the corner of his mouth to his even tinier black and white Friesian calf, which was going all sideways in the way that calves will. Mr Vick, the cattle judge and famous breeder from Steyning in West Sussex, gave the tiny calf's tiny bottom a tiny pat, it walked on and all was well. 'There we are then,' he said, and pulled his cap at an even sharper angle to the horizontal. Wonderful Heathfield Show!

We trade-stand judges were shepherded around by Peter Salter, an elegant man and our steward. He wore a pin-striped suit, a bowler hat and gumboots. For years, he had run the South of England Show at Ardingly. To begin with, we were exaggeratedly courteous about each other's likes and dislikes. Mr Bines liked the way one agricultural equipment merchant had managed to get a combine harvester on to his stand. 'Always some plus marks for a combine,' he said. I said nothing. I liked the way a tent full of little food stalls had the air of 1948 about it, one step up from a village fête. It had that delicious smell of a field inside a canvas hall but Mr Bines did not comment. Mrs Chidson liked the verve of the Sussex Express stand, its honest vulgarity, but neither Mr Bines nor I said very much about that either.

We realized, I think, that we were interested in rather different things and by the time we arrived back at the judges' tent we all, I am sure, had a pretty good idea where the others stood. We sat down at a small round table in the tent, Peter Salter got us each a drink and the horse-trading began. There was the E. Watson & Sons Trophy for agricultural stands of 40 feet and over. No problem there. We all agreed the Young Farmers were outstanding, Agrifactors (Southern) Ltd had made a charming effort and Harper & Eede, who had parked their tractors very smartly in front of their tent, deserved a Highly Commended. We proceeded smoothly to the Percy Meakins Perpetual Challenge Trophy for small agricultural stands. Wealden Smallholdings won hands down, manfully overcoming the local outbreak of fowl pest which had meant they had to replace their poultry display with a Wealden cottage garden at the very last minute. Plumpton Agricultural College came a well-deserved second after they had given us a welcome glass of wine on the way round. Peasridge Livestock Equipment, with a fine display of Equine Dental Chisels, One Step Sheep Shampoo and The Original Bull Shine, enjoyed a Highly Commended in the Percy Meakins.

The judges then turned their attention to the John Harper Memorial Trophy for non-agricultural stands. Some awkwardness set in. Mr Bines was in favour of Parker Building Supplies, which had a small self-contained sewage treatment unit in operation on its stand. I was keen on the food tent that had such a charmingly *ad hoc* 1948 quality to it. Mrs Chidson liked the Hugo Oliver sausages stand. Peter Salter, the steward, asked us, at least, to exclude some from the long list. We did, cutting out both Shell Oil and a man who turned bowls. There we reached an impasse. All we could agree on was the excellence of Hailsham Roadstone's driveway display. I couldn't even remember the sewage unit Mr Bines liked so much and he didn't like my food tent suggestion at all. A slight crackle entered the air. Peter Salter suggested Mrs Chidson and I should have another look at the sewage works. We walked to the other side of the show, had a good look and on the way back to the judges' tent agreed: we couldn't possibly call that the winning stand. The food tent, on the other hand, was precisely the sort of small local enterprise that should be encouraged. Absolutely not, Mr Bines still thought on our return. The impasse remained. 'We're not having Parker Building Supplies,' I said, looking at Mrs Chidson. 'Well, I'm not having Taste of the South-East,' Mr Bines said. We all looked at the tablecloth while onlookers took another drag at the B&H. It was time for statesmanship.

What about, I suggested, giving the prize to Hailsham Roadstone, second to Hugo Oliver sausages and Highly Commended to both Parker Building Supplies and Taste of the South-East? Two Highly Commendeds? Highly unorthodox but in the circumstances the only option. Judicious nods all round. Thank God for that. We could all, at last, get stuck into the G&Ts and the white wines, safe in the knowledge that compromise is always best.

Things were not so well regulated at home. While I was away at the Heathfield show, crisis was erupting at the farm. The handsome-attractive builders we had got in to make the oast-house had run out of money before they had finished the job. We had paid them for the windows, which had been delivered to the site, but they hadn't paid the joiners who had made them. Sarah was at home that afternoon with Rosie, tiny Molly in her arms and Patricia Howie, who had come to help while Molly was still so young.

They were all sitting in the garden on the sunny afternoon when a van pulled into the yard. Two burly men Sarah had never seen before got out, walked over to the oast-house and started loading the window frames, which were stacked against the half-made wall, into the back of their van.

Sarah went over with Molly in her arms, horrified at what

was going on. 'Hang on a minute,' she said, 'those are our windows. What are you doing? We have paid for them.'

'Well, dear,' one of the joiners said, – he was Irish, 'you might have paid for them but we haven't been,' and carried on loading them up. 'We don't like doing this to a lady. But we haven't got a choice.'

Sarah started crying.

Patricia then whispered in her ear, 'I'll distract them. You get the keys.'

'Keys?'

'You get the keys of their van,' Patricia hissed. Then, to the Irishmen, twinkling a little, 'A cup of tea, everybody? Why don't you come in for a moment?' They went inside, and sat down for some tea, while Sarah found an excuse to go out again, took the keys out of the ignition and hid them in a drawer in our bedroom. After the tea, and after they had finished loading the van with the windows, they said goodbye nicely and apologetically, and got in to drive away. No keys.

'I've hidden them,' Sarah said. She had already rung Alex Kelsey, the project manager on the job. 'Get here now,' she told him. It was a Saturday afternoon but he was there in half an hour. He had summoned the main contractors too, and between them all he then held a meeting in the yard while Rosie stood there gazing up at the huddle of enraged men. By the time I returned from the Heathfield show, Sarah and Patricia were in the kitchen, congratulating themselves on a coup; all dust had settled, money had appeared and been transferred, the men had all left, Patricia was the hero of the hour and the windows which I am now looking out through at the spring sunshine were still there, stacked against the wall.

The week after the show has gone down in memory as Chaos Week. My sister arrived at the end of it. She was wandering into the tail end of a disaster sequence out of hell. First, I am afraid, it was the sheep. In the intense and frozen days of mid-March, perhaps a little romantically, I had decided that our lambs didn't need their tails docking. Why should they, poor little things? If God had given them a tail, and so on and so on. So all summer long they had been whisking and flicking their tails – increasingly woolly on the outside and increasingly shit-encrusted on the inside – around the pastures. It was a recipe for disaster, as everyone now tells me. They had until recently done quite well. They were almost as big as their mothers, fat and a little lumbering, but they still gambolled about from time to time, which looked ridiculous, as if Nicholas Soames were playing leap-frog in Parliament Square. You expect a degree of dignity from a sheep and doing hop-skips with a final pirouette of the hind leg, when they should, by rights, already be in some-one's freezer, looks as grotesque as synchronized swimming.

On the other hand, one could perhaps see it as the last gay flicker of childhood before they sank into the morose condition of adult sheepness. Why are grown-up sheep so morose? Why don't they play with each other? It's easy to imagine endless games of British Bulldog, up and down, up and down the fields, until they finally collapse in the evening, exhausted but happy. Why weren't sheep like that?

As it was, at the beginning of that week, four of the lambs suddenly developed the most gruesome condition I have ever seen. It was a case of maggoty bums, and we had to do urgent, heavy remedial work to save the poor things, dressing up in protective gear to administer the dip to their unhappy bottoms, spraying them and anointing them with a pharmacopoeia of sheep treatments.

Sheep are not low-maintenance farming. This has to be understood. You don't get a flock of perfectly whole and lovely-looking sheep on perfectly lovely short-cropped green grass just by putting one on top of the other. That's what I imagined, but it's not true. So, having gone down the wrong track, my sheep now looked as if they'd just come back from a fashion shoot.

They all had blue blobs here and there to identify them as mine, except for those which had green blobs to identify them as Peter Clark's. Those which were nicked in the shearing had patches of violet wool where we had sprayed them with antiseptic. One which had knocked its horn off against a gate-post had a half-yellow head where I had smeared it with a yellow paint-cream which keeps the flies off. The lambs with sore bottoms were now, from behind, a slightly disorientating mixture of Cadbury's-chocolate-wrapper violet and vanilla-ice-cream yellow. I looked at them and felt only a sense of guilt and failure. Perhaps the thing to do was to dip them all in shocking pink and pretend it was on purpose. I had to do better.

Against this background of three-o'clock-in-the-morning remorse and anxiety, there had been a burbling stream of other hopelessnesses. British Telecom wanted to put a giant new telegraph pole right at the end of the garden, dominating Sarah's carefully orchestrated harmonies. So that had to be negotiated away and underground but once it had, I realized the new duct for the telephone wires had been laid wrongly. The reeds in the new sewage system seemed to have died and the smell predicted by Ken Weekes from such a newfangled thing had started to waft up towards the house. A lorry delivering stone had smashed a manhole over the sewage pipe, which didn't help. Then I found the children playing on the gravel among the dead reeds, popping the little pebbles into their mouths in a game which involved getting as many of the sewage-encrusted stones as possible in their mouths at the same time. 'Never let me see you doing that again,' I said, and as I walked away I saw them out of the corner of my eye hunching their shoulders and putting their hands over their mouths in the time-honoured signal of: Uncontrollable Giggles Brought On By Expostulation From Old Fart.

Meanwhile, as the farm and the fields and the stock and I all looked increasingly decrepit and, at times, beyond redemption,

Sarah's garden was entering its glory phase. There was a time when, unkindly, I had referred to that 80- by 40-foot patch of walled, hedged, pathed, manured, sanded, worked-over, reworked-over, planted, replanted, deplanted, weedkilled and cosseted piece of ground simply as 'The Expensive Garden'. I realized now that was a tautology; gardens were £20 notes on stalks.

Even so, the garden had become, in its first real moment of completion, an incredibly beautiful thing, floating free of all that had gone into making it. It was brimming with intense colour, as concentrated as flowers in a vase. It was so full you had to push your way through its paths, getting a soaking in the early morning from the dew, brushing up against things which the sheer height and thickness of the other plants had obscured until you were on top of them, a small colony of sunflowers in one corner, wafting drifts of white and pink cosmos in another, like a hillside in Bhutan.

All week long a stream of garden photographers had been trooping in and out of it for their various magazines. A lady from *Gardeners' World* came to video Sarah talking about the propagation of annuals. Intense tidying up had gone on between each of these visits, but the process had been dogged by something which came to symbolize the difference between garden and farm, Sarah and me.

There was a rogue chicken. It would not go back in the run but instead, just as Sarah had finished her last sweep and survey before Andrew Lawson or Howard Sooley arrived, would wander into the garden and begin to shuffle its way messily through the applied mulch. That was the pattern of the week: chicken, Lawson, chicken, Sooley, chicken, BBC, like some monstrous club sandwich. 'What is it with your animals?' Sarah asked. 'Can't you get anything organized?'

That's when we hit our big sheep crisis. But when wasn't there a sheep crisis? Sheep *are* crisis. The ewes had been having their problems. One of them had somehow cut a tendon in her back leg.

It would not mend and so she had to be put down and buried. Then another ewe started to behave in a way that was most unsheeplike. She would wander off to be on her own, often choosing a place in the field just on the brow of the hill, from which in a dreamy and rather poetic way she would gaze at the radiant colours of the life-burgeoning Weald. This was exciting: a Romantic sheep, evidence of the appreciation of beauty in the lower orders of creation. One evening, the sheep and I even spent some moments together, side by side, looking at the folded view of wood and valley before us.

But I was mistaken. The ewe was growing mad, not wise. I described the symptoms to Carolyn Fieldwick. 'No, Adam,' she said. 'I don't think the ewe is gazing at the Sussex landscape.' The ewe's problem, it turned out, was magnesium deficiency, which brings on a state of dignified but eventually fatal calm. She, too, had to be put down and buried in the corner of a field.

Sheep, contrary to what one might expect, are choosy eaters. They won't touch nettles, thistles, ragwort or dock. What is more difficult to accept is that they will not eat grass – perfectly good, organic, herb-rich, Sussex meadow grass – if it's even slightly too long. This is frustrating. You put them in a field of what looks like the most delicious of salads, and they stand about disconsolately, staring at you with wrinkled contempt, like an aunt who has just stepped in some dog mess. The whole flock reminded me of the faces one sees through the rain-smeared windows of a bus tour of the Scottish Highlands. We didn't have enough sheep to keep the grass short. We needed more sheep.

This was when the crisis began. Our neighbour, Shirley, who had the cottage and a couple of acres on the edge of the Big Wood, was an accountant who worked in the village. She had a few sheep of her own, but they were eating her grass to nothing and one or two had broken out and got into our hay. We wanted more sheep, she wanted to be rid of hers. It was obvious that they should become ours. Here the waters started to become a little murky. I maintained that we had been unable to agree a price; she maintained that we did agree a price, £35 a ewe. Anyway, the sheep were transferred to our fields. The money business, as I thought anyway, was left pending, but we were in dispute about that.

Three of the sheep weren't happy about the transfer. They had been fed ewe nuts at home and, unenthused about the grass-only diet we were offering, decided to go back, breaking their way through our slightly gappy fencing in their bid for freedom. We took them back but they broke for home again and we left them to it.

This would have been all right, but disaster struck. Our neighbour's boyfriend, Dick, a director of a national car business of immense standing, decided to store a luxurious car – a bottle-green Lexus worth £35,000 – at her house to prevent it being vandalized in Heathfield. What safer place could one think of than out here in the Arcadian idyll?

Only days later the poor man arrived at the back door, his face as long as a Blue Leicester tup's. 'Your sheep,' he said, 'have been headbutting my car. They were spotted attacking the doors.'

Have you ever had a phone call mid-morning from your children's headmaster, telling you that there has been an incident at break and do you think you could come over and talk to the parties involved because it is best to sort these things out straight away? Sick in the stomach, I went to inspect the damage. The four doors of the Lexus' were indeed neatly dented at about sheep-head height. The paintwork was buffed up and dust-free at that level too, as if by a fleece still attached to its owner. I thought of suggesting that he might like to keep a sheep in his showroom in Heathfield to maintain his cars in a perfectly shiny condition. I refrained. It was not the right time.

I had no insurance against any damage any sheep on my land might do to anyone else's property. That extra clause to our insurance policy would have cost £30.50 a year. I had thought the idea ridiculous. What would any sheep living here do to anyone else's property? Go and attack it? A night raid on their ewe nuts? It was just another insurance scam. But I pay that premium now.

The sheep/neighbour/limousine/solicitor/insurance company crisis dogged our lives for years. Shirley's son, Jonny, had seen 'my' sheep (whose they actually were remained in dispute for month after month, the meter ticking away in various well-appointed legal offices) attacking the car from an upper window. He wrote a graphic account of how the sheep, in an orchestrated manoeuvre, approached the flanks of the limousine. The synchronized animals attacked the car. They kicked and head-butted both sides, Jonny told the insurance company. Nothing if not systematic. You had to laugh. But then the bill came in. It was £2,300 for repairing the damage to the doors and another £2,000-odd for the loss of income Dick would have had from hiring out the car in the meantime.

Shirley had been away when the incident occurred. Only Jonny had witnessed it, although Dick did admit to having moved some sheep into an ungated field not long before it was said to have happened. We maintained that ewes didn't do that kind of thing. We even paid a professor of animal behaviour from Cambridge University to trawl through the literature on sheep. There was not a single example in the long annals of biological science of ewes attacking cars. Rams have, but never ewes. And ewes have never been seen to kick anything at all.

Despite the power of that evidence, there was no movement on the other side. Their solicitor came down here in his red Audi estate. I walked him around the fields. We followed the presumed route the sheep must have taken. I referred at one point while talking to him, charmingly, smilingly, the sun on our backs, not to 'the sheep' but to 'our sheep'. It was a slip of the tongue. I should have said "our sheep", making that little sign with fingers in the air to show quote marks in the way of American academics discussing "perception" or "reality". But I didn't and the solicitor, with the keys to his Audi bulging in his pocket, and his meter ticking, said, all bright and sharp, "'Our sheep'? You mean they were your sheep, were they? You consider that you owned these sheep at the time, do you?' For God's sake, I thought. Imagine living your life like that.

On it rolled. Relations with Shirley were not good but our lives were painfully enmeshed. We were already in dispute about her water supply and the track to her house. We were like a pair of sumo wrestlers, podgily shoving and clutching at each other. Attempts at settlement and compromise never seemed to work. She became ill. Just when I hoped one of the issues might have faded away, another solicitor's letter or another angry note or dark remark about the track or the water would appear. I offered to pay a third of the track repair costs. That got nowhere. There was some mutual berating. The place where we lived was beautiful but it felt as though it had an abscess in its gum.

On one particularly bad corner in the lane, Anna Cheney collided with a car coming the other way, driven by Shirley. Anna, who looks after our children, had both our daughters in the back of her car. I had a phone call from the woman who lives in the nearest farm and rushed down there. No one was hurt but the girls were in tears, the cars looked mangled and everyone was feeling fluttery and shaken. Except for one of the policemen, who was all smiles and hands-in-pockets, seen-it-all-before, get-this-every-day-of-the-week, what's-the-fuss. I very nearly had a row with him until Anna restrained me, telling me I would get arrested. I realize now it was Shirley I was boilingly angry with.

Two events finally propelled our relationship into the strangest realms. Sarah and I were sitting at home after dinner on a Friday night. The children were all in bed and the dog was lying in front of the fire. Then, as mothers do, even through the chat, Sarah heard Molly crying. 'Sssh,' she said, listening. We heard the sound again, but it wasn't Molly. It was a siren coming up the lane. First one and then another fire engine came up to our garden gate, paused and then went on towards Ken Weekes's house, paused there, saw nothing, and then on again, up towards Barn Farm and Mount Farm, up at the top of the lane.

We sat down again. Some poor family or other had obviously set fire to a chimney. We had done it once a couple of years before. But on that Friday night the fire engines weren't for us. The sirens faded away. After a few minutes, though, we heard them coming back. I went out. 'Where do you want?' I shouted. The first driver shouted the name of a house. It was Shirley's. I told them how to get there, found a torch and ran over there myself.

It was about eleven at night. A west wind which had been blowing all day was still spitting the rain horizontally on to the back of my head. I was only a few minutes behind the fire engines but by the time I got to the house the fire brigade's whole system was up and running: arc lights providing a wash of white light like a film set; hoses unreeled around the house, charged with water and with a junior fireman on the end of each one; a white-board on a tripod where a fireman with a black marker was recording the sequence of events. Two teams with breathing apparatus had gone straight into the building and were fighting the fire in the kitchen. Shirley, still in her nightdress but with a blue anorak over it, was sitting in one of the fire engines, shocked and shaky. She had been watching TV in bed when the electricity had suddenly gone off. She had opened the door to go downstairs to the fusebox, only to be met by a solid wall of smoke, poisonous-tasting, gagging in her throat. The phone was still working and, even though she had to dial in the dark, she had called 999 and then got herself out of the building into the wet night.

The worry was her animals. She had three dogs and three cats. The cats could probably look after themselves but two of the dogs had slipped back inside the house after she had got out herself. They were now trapped in the scullery, between the fire and the locked back door, and she couldn't get at them. She was terrified that they would be burnt or were suffocating in the black smoke.

The first thing the firemen did was to break open a pane in the back door, unlock it, release the animals, all unharmed, and shut them in the stable. The boiler had burst into flames and was still burning in the kitchen. Within a few minutes, though, the crisis was already over and the tension winding down. One crew was coming out of the house, tearing off their oxygen masks, their heads and faces running with sweat. The seat of the fire was out. The other crew was still in there, checking for flames in other parts.

Suddenly, from between the tiles of the roof, and snatched away by the wind, smoke poured into the night air. Firemen shone their torches up at the gables and ran extra hoses round the downwind side of the house. Smoke was crowding out of the roof. The firemen had opened a hatch to the attic, air had poured in there and fire had suddenly erupted, soon taking hold. New urgency gripped the firemen. The fire controller got on the radio and the men in breathing apparatus went back in. Shirley sat in the fire engine. I watched aghast.

The conventional wisdom is that once a fire takes hold in the roof space of a building, it is extremely difficult to prevent the whole roof going or, in some cases, to save the building itself. If no one had been at home that evening, or if it had been a little later and Shirley had been asleep and not noticed the electricity going off, then the chances are that the house would have burned down. For a while no one was sure if this roof fire was going to be contained. There was talk of calling in extra tenders. For perhaps 10 minutes there was uncertainty until,

quite abruptly, the smoke pouring from between the tiles diminished. They had extinguished it. It was 'a good stop'.

I took Shirley and her three dogs home with me that night. We
drank most of a bottle of whisky in our kitchen, waiting for the
firemen to finish clearing up, checking no embers remained.
At about one o'clock in the morning I went back up there. Half
the contents of the attic, old suitcases, boxes of books, the sort of
baggage we all have stuck away in the uninspected corners of our
lives, had been hauled out on to the drive in front of the house.
It was all now scorched and sodden. The boiler itself looked as if
it had been bombed, the kitchen black, wrecked, unusable.

It was the sight of the attic that was most alarming: the rafters
charred, all the implications there of what might have happened,
how near a real catastrophe might have been. A few minutes' more
burning and the end would have been quite different. As it is, or
so says Ken Weekes, who had had a fire a few years before, 'The
smell won't go away. Sometimes you just catch it, just a whiff of
it. It brings it back, I can tell you.'

Shirley stayed with us for a few days, the awkward subjects
untouched, a sort of grinning distance between us. None of us
had any idea what was really going on in her life. There had
been whispers but we had ignored them. She went off to stay
somewhere else, leaving behind a pair of shoes. By chance, just
then, quite suddenly, resolutions: the car insurance company
graciously accepted £3,000 to go away. There had been costs of
more than £1,000 on top of that. I know that £4,000 may seem
like a monstrous amount to get someone off your back but we
were dealing with one of the biggest insurance companies, a rich
and powerful organization, prepared to go to court and spend
who knows how much, to batter us into submission, to extend
and amplify the arguments, to explore the niceties of blame and
responsibility, knowing that our funds would run out before
theirs did. So we paid. Around the same time, Shirley's
ex-husband, a beautifully reasonable man, brokered a deal about

the track; and the discovery of the leak in the water pipe explained the many months of problems over the water bills.

Then came the greatest shock of all. A reporter from the local paper turned up. Shirley had been convicted of stealing from her own clients at Lewes Crown Court and sent to jail for nine months. The reporter also said that over the time we had known her she had attempted suicide twice. There had been murmurings but I had understood nothing. What I had seen as awkwardness and recalcitrance were only the surface symptoms of a life in crisis. She had never divulged the reality. Just a quarter of a mile away across the fields had been someone breaking down, and we had not had the faintest idea. Her house was burnt and empty. The ivy began to grow across the windows. Her two fields turned ragged with thistles and docks.

Despite it all, Perch Hill itself was resilient. It provided resilience. It was in its heart what we wanted it to be. One morning in particular that summer (it was 2 June 1996) felt as if it were the day for which the whole of the rest of the year had been a preparation. The long grind of the winter and its sense of enclosure and endlessness; the seeping way its dankness enters every aspect of your life; the delay of spring, the long poking about looking for spring, many weeks before it has any intention of showing itself – all of the waiting had gone. The day that we had been waiting for was today.

I was up early, having to get some work into the newspaper before people arrived at the office there. As I walked over to my workroom across the yard at a few minutes past five, the tangerine sun had just cleared the upper tips of the oak trees in the Middle Shaw. The ducks and chickens were scratching about on the compost by the old cow shed, there was a big lamb bleating for no reason I could see down in the Long Field, there was dew in the grass and the whole place was suffused with that orange-grey, cold-warm, utterly private light of sunrise.

By mid-morning, the work was done, I'd had breakfast and I'd got the day free. I don't understand how sunshine works but everything that morning looked as if it had acquired another dimension. Far to the east, for twelve miles or so to the hills above Rye, it was so clear that I felt I could see individual trees. Westwards I could surely make out the slats in the sails of the Punnett's Town mill, which is a good hour and a half's walk from here. Was all this simply the sharpness and clarity of rain-washed air? Whatever it was, the whole place looked as a glass of white wine tastes.

I went down to the Slip Field. It is the one field on the farm that we all love best here, and that day it was wearing its midsummer clothes. It is a south-facing bank of about two and a half acres surrounded on all sides by wood, the oak and hazel of the Middle Shaw to the right, the long frondy arms of the ashes in the Ashwood Shaw to the left and, in front of me, at the foot of the hillside, the 2 acres of garlic flowering then in the hazelwood shade of Coombe Wood, a stinking, lush and frothy garden which squeaked as you walked through it at that time of year with the big, rubberized, smelly leaves rubbing up against your shins.

It was the field itself which was the zone of heaven that day. Its slippy soil meant that it had never been reseeded with commercial grass mixtures and so here, between the garlic and the bluebell woods, hidden from the world but open to the sun, was our field of flowers. There were sheets and sheets of the yellow vetch with blood-red tips called eggs and bacon. Here the common blue butterflies flitted in pairs, their blue backs just greying to silver along the outer margins of the wing. Curiously, those precise colours, and their relationship, a silvery lining to an eye-blue wing, was exactly repeated in the speedwells that grew in mats among the yellow vetches. Beyond these beds of eggs and bacon, with a scatter of blue among and above them, where the dog and I were both warmly lying, the buttercups and the daisies, with pink fringes to their flowers, spread out to the margins of the woods where the pyramidal bugles clustered a darker blue against the one or two bluebells that had leaked out into the field. The dyer's greenweed was not yet in flower and only some tiny forget-me-nots and the taller spiky speedwells added to the picture. A holly tree on the edge of the wood had turned pale with its clusters of white flowers.

A slight wind started the field nodding and other butterflies cruised and flickered in. A pale tortoiseshell hung for a minute on the vetches, followed by a bumblebee which pushed its entire body inside the blooms. A big cabbage white flirted with the nettles at the top of the field and then two brown moths, each the size of a fingernail, came dancing in a woven spiral across the hillside, as close in with each other, as bound to and as mobile with each other as the different parts of a guttering flame. The whole wood was needled with birdsong, a clustered shrieking sharpness, inter-rupted only by the jays' coarse squawking, the sudden dropping-off *dwaark* of pheasants and, behind it all, the continuous, laid-back strumming of the woodland bassists, the pigeons in their five-part, broken-backed rhythm, two rising, a pause, two falling, doo-doo, doo, doo-doo, the only soundtrack you need for an English summer.

I was sinking into sleep. The dog already had, and his nostrils were twitching as he snored. There was a drone of light planes. One of Ken Weekes's grandchildren must have been playing football up by the cottage and that snatched-at, childish shouting came in scraps and patches across the fields. A thin but unending river of feathery willow-seeds was blowing from out of the wood, on past me and down towards Bateman's and Burwash. Here and there a thistle standing in the field was covered in the willow fluff it had picked from the passing air.

All this was nothing compared to the soporific warmth of the sun, on the field and on my back. My shirt itself felt hot from the warmth it had absorbed. Even the hot dog next to me smelled nice, but then perhaps I only thought that because he was my dog and I thought him wonderful anyway. I rolled over, turned my face away and down into the grass, buried my nose in the sun-warmed turf, breathed it in, smelled how good it was, its hot vegetable dryness, and knew that coming to live here was the best thing I had ever done.

Peaches on the Cow-Shed Wall

NOTHING IS more calculated to turn one into a pork pie than the arrival of a poet on the doorstep. We had one staying that August and the experience changed me into a no-nonsense member of the Rother Weald Branch of the NFU. 'Don't you realize,' I heard myself saying towards the end of his time, 'that the country is where *food* is produced?' Jason, who was twenty-two, looked back at me with the comfortless gaze of a man who had yet to suffer. He was the son of some people from whom I had shamelessly cadged accommodation and food a decade previously when touring the western half of the United States. Reasonably enough, they had now sent Jason over to do the same to me. We had 10 days of him and then he went off walking somewhere in southern Spain.

Jason was into haiku. It was quite charming to start with and I encouraged him to produce some pieces about the farm here. He gave me his first little Japanese creation on the second day at breakfast:

> Hawk
> twitch of long grass –
> elegance of ambush.

'Thank you, Jason,' I said and felt, at least, that here was someone who was on my wavelength. But I wasn't totally at ease. Jason's wanness, his silences, the way he never said anything unless it came out perfectly rolled and made like a sushi – that was difficult. I longed for him to burp or knock something over or spill gravy on his shirt. Surely a little lack of control might be considered Zen too, mightn't it?

One morning when I came down to breakfast, he said 'Hi, Adam. How are you?'

'Fine, thanks.'

'Good, well done,' Jason said. 'I think it's a great thing for someone to be able to say that about themselves.'

Just as exasperation was setting in, he would come out with something beautiful. I took him to see our sheep. Roger the ram was still in a field of his own, with a couple of ram lambs for company, and the ewes and ewe lambs were all together on the other side of the lane. The mothers were wearing the sheep bells I had bought for them in Majorca. Peter Clark had said the bells were stupid and cruel. 'How would you like it?' had been his unanswerable question.

All the same, as I explained to Jason, I loved listening to them. If you heard them from a field or two away, they sounded like snatches of a conversation which you couldn't quite pick up. I don't know how many times I have sat on a mountain in southern Europe, listening to their hollow, half-carried half-notes. Jason said nothing but that evening he gave me this:

> Sheep bells:
> toc toc –
> olives in the oakwood.

Ken Weekes came to lunch and told his stories all the way through it, beaming away at the gnomic Jason, and Jason played his part to the full, economical in his courtesies and self-contained in his stillness. Ken's performance culminated in his favourite story about the hunt – the one that ends with 'Why

don't you fucking well bugger off out of there?' Jason's silence
was a little black hole at one end of the table. That evening he
gave me this:

> Farmers:
> tang
> of horseradish.

Even though I have since learned that those three lines are a
fairly direct adaptation of a famous haiku about samurai, that
was the high point of the 10 days. For some reason, Jason's offer-
ings became increasingly dark and more obscurely critical, as
though we, and by extension this place, somehow embodied
everything he most hated. Sarah and I would sit up after the
others had gone to bed, wondering what were the implications
of one or two of them:

> Perfected garden:
> the end
> of youth.

which did nothing but fill us with gloom. Or there was:

> August sunshine:
> old brocade
> hanging from the trees.

That sort of thing was alarmingly exact, identifying precisely the
upholstered, claustral thickness that gathers around this time of
year, a sort of outdoor stuffiness, which on the worst of days
affects everyone's mood. I came to think that a general air of
discontent was being spread through the entire household by
this man. He came to seem more and more like a parasite, indifferent to the difficulties of getting on in the world and with

one's life. And there he was, eating my food and drinking my drink. I saw in him, I now realise, what some of the serious farmers round here have said they see in me: an exaggerated aestheticism, an ignorance and even arrogance about the facts of life. He was the kind of fantasist who is so bound up in his fantasy that he doesn't know that it is one.

But he couldn't be so casually dismissed. What he said did have an odd and unnerving access to the truth.

The morning he left, he gave me his final verdict:

> Sheep bells:
> toc toc –
> empty gestures.

I folded the piece of paper twice, into a little square, and put it in the bin, feeling hollow and old.

After he had gone and for years afterwards, what he had said, or more what he had suggested, often came back to me. Where really did I stand? Was I just an aesthete, interested in no more than prettiness? Was I really just an empty-gesture merchant?

I don't think so. And I pin my justification to this: increasingly when people arrived at Perch Hill they responded in a way I came to recognize. They saw, first, that they weren't coming to 'a country house' in that gravel-and-Labrador style. This was not a place in a tweed suit. Its ragged edges seemed like the signs of vitality, not neglect. Just as a busy workshop or a sculptor's studio or a writer's workroom always has an air, at least around the edges, of near chaos, the tidelines of past projects laid one on top of another, as if the meaning of the place is not in its finished effect but in the making itself, in sparks flying when live wires are joined, or that wonderful hiss as the iron is quenched in the deep black water tank in a farrier's workshop – maybe these analogies are too strong but Perch Hill seemed to these people to be beautiful in the same busy, intent way.

The place, then, was almost a side-effect of the life that was lived in it. When people came here, they noticed not a place but a life, something under way, with all the chance of failure and raggedness and a mismatch of resources to ambition which that involves. Off-chance visitors tended to stay a little longer than they had meant to; they came back with their friends; they walked around the garden and farm with the sort of look on their faces which they might have when tasting a new kind of pudding or wine: curious, quiet, silently delighted, wondering what it was they had stumbled into here. I feel sure the reaction came from an understanding that in doing what we have done here, we have allowed it to breathe. So much was there before we came and we have given that new expression. We have recognized that we are dependent, not dominant here. And that look on the curious, bemused visitor's face is what I would show anyone who said I was a dilettante, a poseur or an empty-gesture merchant. Like millions of people all over the world since human beings first gave up hunter-gathering and made for themselves a place they liked to think of as their own, we have pursued and nurtured something good here, which is about settlement, sustenance and steadiness shot through with vitality, beauty and delight. And what is trivial about that?

There were some strangely late peaches that September growing against our cow-shed wall. By the middle of the month most of them were ripe, if still rather small, the size of apricots. Those autumn mornings, their skins were wet with dew. One day, Rosie and I ate them for an autumn breakfast, standing in the cold morning sun just after eight o'clock, me with two jerseys on and her in her school uniform, side by side in the vegetable garden. The sun was coming over Coombe Wood and the juice dribbled down our chins. It reminded me of the story of James Thompson, the luxurious 18th-century poet of *The Seasons,* who was found one morning in the fruit garden of some large country house, nibbling at a peach that was still hanging from the tree. He had his hands in his pockets. What a sight: those lips and teeth gobbling to catch the mousy skin, the fat poet, his unbuttoned stance, his casual acceptance of paradise dangled by his nose. It was like a little model of what Perch Hill was giving us: nothing huge, nothing lush, but almost unbearably sweet.

The autumn storms came bulling into our lives. One night in mid-September, I felt as if the house had been at sea all night, the frame of the wooden roof above our bed adrift in a wrestling wind. I thought I must have been dreaming it, this manhandling of the building by the air, but I wasn't. In the morning, the frame had clearly shifted. Hairline cracks in the plaster had opened a hair or two wider. Everywhere you looked you saw them: dark, roughly drawn graphs up in the corners of the rooms or where the roof met the walls, little jagged longitudinal lips that had opened and only half closed, like the caulked seams of a ship that had worked in a storm. If you put your cheek next to one of them in the bathroom, you could feel the air whistling in. This was what these wooden houses did. All our rooms upstairs had the marks of old cracks that had opened and been sealed, opened and been sealed, over and over again ever since the frame was put up 400 years or so before. Ken Weekes said it was in the wood. 'That's one thing you can be sure of,' he told me. 'Oak never dies.'

The wind was still careering around the farm. I was dealing with my agonized tax affairs and no weather could have been better suited than those equinoctial gales. It went without saying, of course, that the staff of the Inland Revenue South East office in Cavendish House, Castle St, Hastings were some of the most charming and helpful on the planet, and were preternaturally understanding about cash-flow difficulties and revenue-stream bottle-necks (my ref: 057 D 58979/NCP). Nevertheless, those 'Dear Mr Nicolson' letters did have something of the reality dose about them. It was all the oast-house's fault, its massive budget overrun, knocking out cold any chance we might have had of staying out of a debt even deeper than the one we were in already. I was working for newspapers at the time, writing anything for which they would pay me, columns, features, interviews, op-ed pieces, straightforward reporting, thousands of words pouring out week after week and the income sluicing down the sinkhole called Perch Hill Farm. At the time Sarah, looking after our little daughters, was working mostly in the garden and not earning much. The whole edifice was teetering on the rather narrow basis of the computer on my desk.

The tax demand sitting on the desk in front of me that morning was a fat, slapping, smacking squall of reality coming snorting into our lives. All those balmy midsummer ideas for this and that change, this fencing and hedging scheme, that alteration to the back of the house, the new roof to the barn, the new tractor and cart shed: the whole lot was blown heedlessly away. I could stand at the window of the oast-house and watch our plans being tumbled and driven far out to the other side of the millennium, those distant, sunny, cash-filled fields, from which the burden of backlog tax, a pernicious weed, would at last have been removed.

There was an exhilaration to this moment. At least the thing had been faced and we were still there. I had been walking around the farm that morning, in the wind, the real wind I mean, trying to chew over schemes for generating some money and delaying the payment of bills, but I could hardly concentrate. My whole field of vision was taken up with the excitement of the wind's sudden, savage blanching of the trees. Summer was over; here was the gateway to a whole new way of seeing things.

I'd never looked at wind the way I saw it that morning. It articulates the landscape it disturbs. All still things are alike in their stillness but every windy thing is windy in its own way. That's the stripping pleasure of a wind. It breaks the landscape apart; each element becomes itself. Perhaps this was no more than a mechanical thing, the angles at which branches join the trunk, the inherent strength of twig and leaf, but the effect is that the wind makes the landscape talk.

Outside the garden gate the willows were driven by the wind, their bright and glaucous underleaves exposed as the wind blew down on to them. They were like girls with blown hair. I looked at them and saw an actress, a long-haired blonde – is it Julie Christie? – her hair blown hopelessly all over her face. She picks at strands of it but the wind plasters them back across her cheeks and eyes. She shakes her head to free herself from them but they won't leave her.

In the garden, the Victoria plum was still overladen. We should have unburdened the tree of the fruit but we hadn't and the top branch had snapped with the weight. Poor old thing, bowed down with her trappings, not Julie Christie but a jewelled duchess, somehow, tragically, or at least pathetically, broken down mid-ball.

There was a general cowering of the trees. The whole roof of the wood, as I looked out across the five miles of the Dudwell valley below us, was glossed and whitened in the wind. The blasts were humpbacked as they arrived at us: a low and slow building to a peak in which the big oaks suddenly seemed much bigger, out of phase with themselves, a broken swell stirring about in their branches, with different winds in different parts of the tree and then, quite suddenly, the gust dropping sharply away and we were waiting for the next.

The clouds cruised past the tiled ridges, a frame a second, and in the distance a deeper grumbling note marked the passage of a gust through the big wood. Down in there, sheltered by itself, the wood was still. The upper branches were waving among themselves and it was like standing as a child in a giant football crowd, with a crowd of waving men far above you, that sweeping-dying roar, but hoarser than a roar, almost a whispered roaring,

Years of topping the thistles and using no sprays
on the pastures have produced a rich, flowery
and diverse sward full of vetches and clovers,
as here in the bottom of Great Flemings.

Bluebells and orchids in the wood.

The Perch Hill herd of Sussex cattle in Target Field: as soon as they arrived it felt as if the farm was complete.

Glossy, full and fat on Perch Hill grass, the cattle came to seem what Perch Hill was for.

Simon Bishop, with Norman the Sussex bull and one of his cows, steered the farm at Perch Hill into a wonderful, new and healthy state which it would never have reached without him. He was tragically killed in a road accident in November 2009.

California and Iceland poppies,
cornflowers and foxgloves fill a misty
summer morning in the cutting garden,
as the house now sits in its sea of flowers.

Beech Meadow, which in the 50s and 60s was used as the farm's only arable field, now alternates between pasture for sheep and cattle and a hay meadow.

White foxgloves line the hedge between the fruit garden and the entrance track. A standard hawthorn planted in 1997 leans away from the prevailing southwesterlies.

The converted cowshed now has a
green house attached and is surrounded
by a mixture of artichokes and herbs.
This has become Sarah's garden and
cookery school.

Sarah talks to a group of visitors
in the new vegetable garden on
the virtues of growing your own.

The smell of summer grass.

heard far above you where, in manic circular motion, the trees were waving at events you could not see.

Out on the other side of the wood, the blades of new grass, after hay, twitched and flickered, the antennae-whiskers of the meadow. Along the edge of the Way Field, the hollies clicked like the quills of an anxious bird. The seed-cases of the hornbeams beside them made the high, dry chatter of rattled maracas. The ashes were all blown to pieces, their fronds offering no more resistance than a palm in a hurricane. The beeches, hammered by the summer, were now hammered again by the first of the autumn wind. They looked more bruised than anything else but at least the lower, sheltered branches were all right, dark but still fresh. It was up in the windy fringes of the beech that the leaves suffered, blackened in the wind, desiccated and made useless. No tree travels a greater distance from bright to dark in the course of a summer, from maidenly to matronly, the spring's most brilliant debby green to this heavy darkness, a whole life in a year. Beneath them, on the small birches colonizing High Wood, every little leaf was struggling on its imprisoning stem.

There was a sort of bustling, a boxer's movement, to all the big trees here, the oaks especially, working at the wind, weaving away from its advances, a knitted, dodging, evasive dislike in the movement. The wind was unkind, the end of summer: we'd had the cheque; here was the bill.

Some money had to be raised and so we decided to take some of the old ewes to market. I went through the flock with Fred and Margaret Groombridge, picking out the ones that were past their best and, as Margaret said, 'would only be trouble if you kept them'. The ones to go, thirteen of them, were marked with a squirt of green spray paint on the back of their necks.

The Groombridges brought their trailer, and we hitched it on to my Land Rover, loaded up the old things and drove them into Hailsham. The sun dabbed at the green tunnelled lanes. I drove at 40, with gradual stops and cautious starts, conscious of the animals packed in behind me. Margaret filled in the Sheep Movement Declaration Form: Holding number 41/019/0216; total number of animals: 13; identification mark: green 'A' on flank.

We unloaded the sheep into holding pens in the market and they were sorted into three lots: half good, not so good and unspeakably bad. Other sheep from other trailers were going through the same process and, once sorted, were all then driven through the maze of hurdles and gated corridors into the selling pens.

I have heard that markets put heavy stress on the animals that go through them and that cattle, for example, cannot be expected to grow for a month after they have been to market, but it did not seem too bad for the sheep. All the men who handled them talked to them as they did so – 'Come on old girl, come on then, this way then' – and it wasn't rough.

Even so, to my own surprise, I felt a little cheerless about it all. Some of these old sheep were from the first lot that I'd had, and looking at them now, in these market pens, I couldn't help but think of that day when they had first arrived here and we had led them all out of the trailer and into the Long Field. Everyone had felt that day to be a step forward here. 'Well, well,' Ken Weekes had said, 'livestock back on Perch Hill Farm.'

Memory summoned other times: lying out one night that summer in their field and in the dark finding, amazingly, that the sheep luminesced – round, moonlit bundles scattered across the grass; this spring's lambing, the anxiety, the acid cold of the March nights and the strange, casual passion of it; one or two of their sisters dying, for no known reason, in the middle of an otherwise eventless summer day. 'Animals are the soul of a land-scape,' Rudolf Steiner said, and in Hailsham Market that was the phrase that came to mind.

There were other things there too. I didn't want these creatures to go for nothing. I wanted a good price. The auctioneer and his cluster of buyers were starting on the fat lambs down at the other end of the market and the rhythmic, lulling-aggressive banter of the auctioneer came drifting up over the pens. The old ewes, the cull ewes, were the last category to be sold and it so happened that my three pens were the last of all.

There were three big buyers there, one a wholesale butcher from Essex who buys 140,000 cull ewes a year for what is called 'the ethnic trade' – Muslim retail butchers and Indian restaurants, mostly in London, doner kebabs, mutton curry, that sort of thing – and a couple of Sussex farmers, who will pick the good from the bad and make their profit that way. The three were all physically big men. The Essex butcher was shaven-headed, had two pipes in his back pocket, a mobile phone attached to his waist, and drove a shiny new Discovery. The joshing between them was tough, commercial, competitive, no more than half-funny.

I stood and watched as the noisy cluster came near our pens. 'Straight off the Downs, they are,' the auctioneer said of one lot, the ends of his tie tucked in between the buttons of his shirt. 'Real good sort of big things, they are. Don't be stupid. Has she got teeth? She's got more teeth than Hannibal Lecter.' The sheep-dogs were nosing about the alleyways. The auctioneer described one of the buyers as 'the onl fifty-five-year-old bachelor in Sussex still living with his Mum. Aaaaah.' The man flicked him a V sign and everyone laughed. Prices were running high. There was a sort of savage buoyancy in the air.

The group arrived at our pens. 'From Brightling,' the auctioneer said curiously, 'Mr Nicolson.' I put my hand in the air and the small group turned to look. 'Some big Suffolks,' the auctioneer said and the serious buyers, with their sleeves rolled up over hairy arms, climbed into the pen and felt the backs of my sheep. A muddle of protective-defensive thoughts went through my head. 'Twenty-eight, eight-fifty, nine, nine-fifty, thirty I have.' The price climbed, quickly, dispassionately. 'Are you bidding or are you having an itch?' He was bidding and the sheep went in the end to the Essex butcher for £35 each. The second-best three also went to him for £33 each. We came to the remaining seven, the truly bad, end-of-their-lives animals, and the Essex butcher turned away. They were not worth the bother, and he pulled some tobacco from his pocket. The auctioneer squeezed a bid of £12 out of one of the Sussex farmers. There was no energy in it, the lot was an embarrassment but somehow, still, the price crept up and these old scrags were finally knocked down for £18.50 each. 'Stupid,' Fred Groombridge said. 'I wouldn't give a fiver for one of them.' So it had been a great morning: £333.50 (minus commission) for thirteen end-of-the-road sheep, all thanks to BSE and the flight from beef. Watching them walk up with the others into the big red, multi-tiered Essex lorry, the green As still there on their sides, I felt as if my children had done well at school. The money would pay for something; not much, but something.

At seven o'clock the following evening, I was feeding the dog when Ken Weekes's burglar alarm went off. Sarah and I knew what it was: the manic oscillations of the alarm, its unworldly wailings, shrieking and booming across the orchard and the Cottage Field. The early night was already solid black, it was raining slightly and I didn't think twice. I shouted at the dog to come, grabbed our giant Dragon torch, a sort of portable headlamp, and ran out of the door.

I had to hurry. Ken had been burgled in the daytime a few weeks previously and on that occasion I was up at his house within about five minutes of the alarm going off, to find the front door kicked in, the jambs smashed to pieces and a drawer in a bedroom upstairs hanging half open and rifled, like the lolling tongue on an exhausted dog. I then crept around the house, banging open the door of each room as I came to it – this felt absurd, cop showy – anxious not to find a person inside, but the burglar had gone. Later that day, Ken's wife, Brenda, found one or two of her things scattered along the lane to Burwash: an old bundle of tissues, a pendant, treasured but not valuable, thrown from the burglar's car window like the discarded wrappers from a box of chocolates.

So I knew I had to hurry if I was going to find anything more substantial this time than a broken pane of glass and an empty house. The beam from the torch was jagging and veering all over the lane and the wood as I ran uphill. I had passed the gate into the Cottage Field and was opposite the entrance to Blackbrooks Wood when I saw the man coming down the road towards me. His whole body was white in the beam of the torch. The white drops of rain were flicking everywhere around him. He was coming definitely and deliberately towards me.

That was frightening; there was no fear or reluctance in his movement, just a steady walk down the lane towards me. I had stopped as soon as I saw him and, as he kept moving towards me, my own fear grew. I kept the bright beam of the torch on his face, still 80 yards away, and screamed, 'Who are you? Who are you?' He didn't answer but kept on towards me. I shouted at the dog to get near me and began involuntarily to walk backwards, the torch still on the man, keeping the distance between us unchanged. He was holding something in his right hand. 'What's that in your hand?' I screamed at him. 'A bottle,' he said. 'A water bottle. I know this looks bad.'

I felt tight in my neck. This stupid situation, this chance of a fight in the lane at night in the rain. I heard my own voice coming out desperate and violent. 'What the hell are you doing here? Don't get any nearer me. Stay there. Stay there. Stay where you are, will you?' This is simply how fear makes one talk.

All he was getting was the beam of the torch and the strangulated, wildly aggressive voice in the dark. He was now 20 yards away. He had an empty plastic Evian bottle in his hand and he was holding his head away from the light in his eyes.

He had trainers on and an anorak. 'I know this looks bad,' he said again.

'Get down there,' I said. 'Walk in front of me. Get down the fucking lane.'

I let him pass me on the far side of the lane and we started walking back down towards our house. I kept 10 yards behind him and as he walked in the beam of the torch in front of me, he put his hands in the air, surrendering, his right hand still holding the empty water bottle, dangling up there in the light of the torch like a lit and rather blurry lantern. I hadn't asked him to do this. He was treating the torch as though it were a gun and I can only suppose that my own terrified voice had in some way terrified him.

Of course, I hadn't worked out what would happen next. What if this man had been violent? What if he had been armed? What if he had tried to knife me? Not a single moment's reflection or calculation had been given to those questions.

Still shaking, I marched him into the kitchen and told him to sit down. I saw then that he was shaking too. Sarah called the police. I offered him a cup of tea, which he refused, and I then wrote down his name and address. He said he'd been out jogging. He had a whole string of emotional and financial crises in his life at home. He knew this looked bad. It was the last thing he needed. He looked at the floor and rubbed his head and eyes. I didn't ask him any questions, I didn't search him and I didn't feel, once he was sitting here at the kitchen table, that he was any threat.

The police arrived, in two cars, a policewoman and two men. They had taken 20 minutes from Hastings, fifteen miles away on small country roads. Two of them searched Ken's house for signs of break-in and one sat in our kitchen talking to my man. 'You look nervous,' the policeman said. 'Frightened of what they're going to find?' He said he wasn't, and he was right. The searching police returned: no hint of a break-in. It was a false alarm. 'There has never been a false alarm with that system in two years,' I said. 'Must have been freak weather conditions,' the policewoman said. They drove the man home and I comforted Rosie, who had been crying with the commotion. Next morning Ken came over and told me that the detectors on the windows were sensitively set. 'You've only got to tap the glass and they'll go off,' he said.

Sometimes in the summer Sarah and I have lain awake at three or four in the morning listening to the poachers in Blackbrooks Wood, on the other side of the lane from our house. The stinging smack of a rifle shot in the middle of the night is a strange thing. You feel as alert as the deer for which it is intended. You listen, entirely awake, for the sound of movement or even voices over there in the wood. There is silence and you wonder . . . That silence lasts and then, from nowhere, the noise of an engine pulls out into the night, the headlights stream across the dark, the smoke of the diesel exhaust is picked up in their beam, and then the van turns away, its lights throwing the hedgerow trees into silhouette, and you listen as it moves off towards the village.

It is the sort of thing that makes you wary, much warier than either of us would like to be. Ours was the only house in the lane that was not burgled in the first two years of our being there. In the past, this has been classic thieving ground. As Roger Wells quotes in *Victorian Village*, his study of nineteenth century Burwash, it was said of this parish in the early years of the century that the 'labouring Class had become very dissatisfied, disrespectful and insolent to their superiors, riotous and turbulent, ready for extreme acts of depredation, prone to Robbery, violence, and lawlessness'.

Only the truly rabid would think of describing Burwash now in quite such fruity terms. There is no more than a little burglary here and there. Nevertheless, there is no need to be starry-eyed about the facts. Despite the apparent slosh of money, we live on the edge of an area of high unemployment (eleven per cent, compared with 6 per cent further north and west around Tunbridge Wells or East Grinstead), which has been given special economic status as a Rural Development Area. Poverty is an everyday reality. So, here on the boundary of rich and poor, with what looks like disposable wealth lying about unprotected outside, theft seems an inevitability.

There is one form of wealth readily to hand and it is regularly taken: roofs. Roof theft is one of the signs of our attachment to a hand-made past. You are driving along a lane. It's one you know well. Just around the corner, as you come to a rise from which you get a sudden view over the Weald, there is an old farmhouse with a huge pitched tent of a roof.

The roof is a wonderfully crinkled micro-landscape of its own, covered in clay peg-tiles, many of them crusted in yellow lichen. Some of the tiles are bowed slightly out of true so that the tiny shadows thrown by their lips vary as your eye runs along the length of them. These Wealden roofs are beautiful things, pegging the houses down into the landscape. They are often, in this steep up-and-down country, the first part you see of a house or its assembled barns.

But one day you turn the corner and the roof isn't there. Or at least the roof structure is, the timbers, the felting and the laths for the tiles, but the tiles themselves have gone, ripped off and removed. The house looks odd, top-light, so to speak, like a guardsman with his bearskin off. Usually the tile thieves don't manage to get every last tile and the roof is left with quadrants of its old surface in the uppermost corners. The arc of what remains is a measure of the thief's final reach as he stood on the roof-battens before a car in the lane panicked him or his co-burglar thought they'd got enough and to risk any more would be dumb.

Roof-thievery is profitable enough. You can't exactly lock a roof up, and for a thief it will represent a good night's haul. A fair to middling roof might have 12,000 tiles on it. If you were to buy them from a builders' merchant, they would cost 70p each or more. The merchant puts on a hefty mark-up; anyone selling tiles ('from a little barn this friend of mine in Hampshire wanted getting rid of – he wanted to put a nice garage up') will be lucky to sell them at 45p each, £450 a thousand, which would be £5,400 for the load.

I am told that if there were a pair of you, and if you really went for it, you could get 1,000 off in half an hour, that's six hours for the roof, in the depth of the night, 11 p.m. to 5 a.m., and then you are away, the yield as anonymous as you like. One tile looks pretty much like another. There are, of course, colour variations between tiles made within a few miles of each other, depending on the clay, but that makes passing the goods on even easier. Identifying the area from which tiles have come is nearly impossible. It's a sure-fire theft. Someone somewhere will, soon enough, be happy to pay the owner of an antique architectural materials showroom something like £8,000 for your night's harvest.

So the wheel goes round: demand for old-looking roofs means, inevitably, the destruction of old-looking roofs. I know one new roof on a building a few miles from here that was completed on a Friday and had disappeared by the Monday. Everyone around here now has insurance against tile theft and you could look on the whole business as quite a contribution to the local economy. I've heard it said that roofers, short of work, have stolen the tiles, sold them, advertised their services to the recently deroofed and then been paid to replace the tiles they had so recently lifted.

At the approach of winter when the world prepares for closure and withdrawal into itself, you come to appreciate the enclosure of a house, its protective envelope. This was the time for shutting down, for the dumping of excess and the reduction of risk. Old leaves padded the sides of the lane so that you found yourself driving down a brown leaf runnel. The sodden bark of the trees and their twigs were leather-jacketed against the outside. The ponds were full to overbrimming. No insects moved, except one morning a lost butterfly, flittering inside my workroom and dying there, so that it came to lie stiff and dusty on my desk like a fragment of an old dress too fragile to wear. The hops I had hung up in the autumn turned crisp with the heating. A bee, drunk with winter, crawled hopelessly across the window, all wrong, bemused in a nectarless world.

At night with the torch, that November, I could catch the amber eyes of the deer, grazing out in the field beyond the wood now that the leaves had fallen. The grass was the only food to be had and the deer were dining on it in secret. In the early morning, when the light was still hesitant, half there, with the arms of mist pulling back into the wood, you could see the deer's dark bodies, still grazing, the last of the night shift before they too faded back between the trees, present one minute, absent the next, as the owls hooted for the last time and the day came on hard, if somehow still dark for all the light it brought.

All day long the fox would cruise up and down the fields between us and Coombe Wood. He was furtive and cat-like, a sneak among the grass, so that at times he disappeared behind a higher tuft, a slink of reddish brown in the dew-soaked field, low-slung, white-eared, padding through his territory. It was the time to worry for our ducks and chickens, which had been wandering about all summer and autumn as though no threat existed in their lives. At least the ducks had an island in the pond. We were making a new ark for the chickens around which, no doubt, the fox would taunt them at night, round and round with the smell of fear, the brilliant killer and his stupid prey. But what would happen when the hard frosts came? What would happen to the ducks then? The pond would be no moat for them. I would have to shut them up for days on end. It was not the time for openness.

I felt some hibernating impulse at work in us too. I found myself wanting everything tidied away, as though all the looseness and excrescence of the growing time of year were inappropriate now. We had been messy all year. Old cement sacks went wafting around the yard. Piles of semi-dealt-with stones, old paving slabs, some hardcore which never went into the trench it was intended to fill, bricks for a retaining wall which had yet to be built, steaming piles of manure, bean-sticks pulled out once the beans had been picked – all this lay around like the remains of a party no one had bothered to clear up. Urns from some flower arrangement Sarah had done in July were still where she had dumped them out of the Land Rover one warm summer evening. The lobster pot I had brought back from the Hebrides in early September was still stuffed into a corner of the cow shed. None of it looked fitting any more.

At least the fields were good and tight. This year we had, for the first time, managed them in the right way: grazed them hard over the winter, right down to the bones of clay and mud, let the best of them grow away to a big hay crop which we took in mid-July and topped them twice after that. They looked as grassland should and I didn't mind taking people round. 'This field looks nice,' they said. What better feeling could there be than that, a well-made, well-kept piece of landscape for which all the people here who had helped over the year were responsible. Will and Peter Clark, Dave and Carolyn Fieldwick, Ken Weekes, Fred and Margaret Groombridge, Ray Bowley: they all had a share in how ready the place now looked.

We had sold our ram lambs and got good prices. We had sold our old ewes and old Roger. The remaining sheep were now with the new ram, a big young Kent, and would be lambing with his progeny in April. We had remade the driveway and redug the pond, putting it back where it was marked on maps before the war. An old man came by one day and said, 'I'm glad to see the pond back.' That was good. The oast-house had been finished. The trees planted in the orchard last winter seemed to be growing all right, although one or two had died and needed replacing. The financial crisis of the autumn had been weathered, eased away. I had signed a contract to write a book, some money was coming in, our heads were above water. We could look forward again. There were hedges to go back on the old lines from where they had been torn out over the last 50 years or so. I was thinking of planting a new wood to divide up our biggest field, Great Flemings, and to give us chestnut poles for future fencing and hazel for the wattle windbreaks without which gardening up here was nearly impossible.

Sarah wanted to grow organic vegetables on a commercial scale and we drew out the plot on Beech Meadow, the most fertile of the land here. (Although we never made it.) As the year closed down, and as the dark began to colonize the beginning and the end of each day, the tightness which that brought seemed enabling not debilitating. Hibernation was clarifying the purpose of being here at all. The rush and tumble of the more open times of year tended to obscure the point, which now, in the growing dark, emerged quite clearly. It was this: simply to get it right, to do it properly, to make something good that integrated all the virtues of a good landscape, a good place for people to be and work in, and a good way of farming and gardening. For the first time, on these cold and sodden mornings, I was starting to think we might one day get there.

I started sowing yellow rattle seed in Beech Meadow. Because days were short and I was too busy, I had been doing it just in the dark of the evening, as the frost was coming on and the grass began to crunch underfoot.

Yellow rattle is a slight and unimportant plant, which grows in hay meadows that have not been improved with new thick grasses and clovers. It is a sign of infertility, and infertility is, on the whole, what wild flowers like. At the end of summer, in the hay, the yellow rattle develops little round dry pods, which is why the plant is known, at least round here, as yellow bollocks or rattle bollocks, the loose seeds inside the pods still rattling even when the hay has been baled up.

But there's something else about yellow rattle, which was the reason I was walking slowly up and down our fields with a bag full of the crusty seed those winter evenings. It was more than just a means of escaping from an overheated house. Yellow rattle is semi-parasitic on grass. If you can manage to establish it in the sward, it will begin to drain away some of the vigour with which the dominant grasses grow. It's a way of reducing their dominance, of beginning to provide the conditions for some kind of herbaceous democracy in which all those other flowers we want here, the dyer's greenweed, the vetches and the orchids, might begin to decorate this landscape again.

It was only a trial patch of three acres. I bought three pounds of the seed, a pound an acre, for £60. It had been collected the previous summer as part of a pilot project to restore floweriness to the meadows of the High Weald. No one was quite sure yet whether it would work, but the large plastic bags in which the seed came looked like the promise of a diverse and beautiful future.

I walked up and down in the cold of the evening, on the grass grazed tight now by the sheep. We would keep them in there until the end of February and then shut the fields up and let everything in them grow on until the hay was cut, late in the summer. I took handfuls of the seed out of the bag as I walked along, letting it dribble out between the fingers of my hand. I didn't scatter it as I might have been tempted to, like Millet's Sower, flinging handfuls of the precious stuff into the last of the evening light. You need a far more even, as if windblown, distribution to get the maximum effect.

Up and down, up and down across my fields, I felt like a draught-horse at work, not really considering what I was doing but in a removed and contemplative frame of mind. The clear plastic bag in which the seed came was identical to the clear plastic bag – albeit inside a small copper urn – in which I and my sisters had taken our mother's ashes a few years ago to a place in Switzerland she knew, a flowery meadow above Wengen in the Bernese Oberland, where, before she died, she had asked for them to be scattered.

This sowing of gritty mixed seed was rather like that strange, half-sad, half-gauche moment. I didn't dribble my mother's ashes through my hands. I simply tipped them slowly on to the grass straight out of the bag as we walked along. The wind took the lighter stuff away but the bulk of it fell on to the Alpine meadow like a top-dressing of some kind. I wondered then what the effect would be on the plants, what the biochemistry was of human ashes on Alpine flowers. Should I aim to spread them more widely? Would too much in one place be too rich for whatever was living there, too many trace elements in one helping? I started to swing the mouth of the bag to and fro, scattering the ashes more widely and thinly in the way that the nozzle at the back of a lorry salting the road scatters the salt in even swathes across the whole width of it.

At the very end of that bag, as of this one with the yellow rattle in it, the same moment came. Some specks of grit remained stuck in the corners, somehow held there by the plastic, so that only by holding the bag upside down and tautening and snapping it along its base would those few flecks of human ash, or, now, of yellow rattle seed, drop out on to the grass.

Why was it so important to leave these bags pristine and empty? Why must even the slightest speck be distributed on to the ground? Because ash and seed are not to be wasted? Surely not. There was no practical consideration here. But the idea, then, of putting even a microscopic fragment of my mother's body in the waste-paper basket in a hotel room in Grindelwald – which is where the plastic bag went eventually, along with the urn, so flimsy that it bent like a Coke tin – was unconscionable. And perhaps, with this seed, the memory was too strong and the parallels too close to treat the yellow rattle in any other way.

One bag and one sowing felt, however odd this might sound, like a continuation of the other. And when I saw Beech Meadow washed with the pale flowers of yellow rattle the following summer, I thought of the hillside above Wengen, from which the hay had recently been cut and raked, and where the tourists on the mountain railway were looking out through their passing carriage windows, wondering what my sisters and I were doing with that bag of something, halfway through a rather sultry morning one weekday in July. And of course, in part at least, we were wondering the same thing too.

A World in Transition

Everywhere around us there were signs of a world in transition. One evening I went to a meeting at the Horseshoe Inn in Windmill Hill. Farmers and their wives, perhaps 200 of them, had come here – from the High Weald, from the Pevensey Levels and from the Downs. Here were the Sussex farmers, not rich men, nor in your tailored tweeds, but in heavy jerseys drawn tight over the shoulders, rough, kitchen haircuts, faces in which the wind had broken the veins, and a certain tense reticence, a lack of ease in the public forum, overlying a sense of unfairness, of an injustice being done to them.

They were all stock men, cattle and sheep farmers, and they all used Hailsham Market to buy and sell. I'd seen some of them there before. It was a necessary part of their business, to realize some cash when they needed it, to sell at high prices and buy at low. The nearest other markets, apart from Rye, which was small, were at Ashford in east Kent or Guildford in Surrey, both possible on a big occasion but too far on a regular basis, taking up most of a day to get there and back. These small farmers needed Hailsham Market if their businesses were to work.

But Hailsham Market, like nearly everything else in their lives, was under threat. The government had banned the building of out-of-town supermarkets and so the livestock market's location, within the confines of Hailsham itself, made it an ideal candidate for development. No one was sure who wanted it, but the name on everyone's lips was Sainsbury's.

The company that owned the market was getting £24,000 a year from the site. If they could sell it, with planning permission for a supermarket, it might be worth £1.5 to £2 million. In other words, the current return was 1 per cent of the site's potential value. If the market could be closed, the site could be sold, the shareholders would realize a huge amount of money, the super-market chain would be happy . . . and East Sussex would have lost its only cattle and sheep market.

In a back room of the Horseshoe Inn, a grey, functional, modern space for company dinners and anniversary dances, the farmers were crammed in, standing against the walls when the chairs were full. The local press sat at a table on the side. We were addressed by the chairman of an action group and then by a solicitor who described the threat to the market and the way to lobby against it.

Already Wadhurst, Lewes, Heathfield and Haywards Heath had lost their livestock markets to the same pressures: a growth in direct sales from farm to abattoir and an increase in the value of the market sites for other purposes, such as supermarkets or housing.

The solicitor, in his beard and grey suit, his fingers interlaced on the desk in front of him, dispassionately described what was obviously felt as an emotional issue in the room. Although his purpose was to motivate the farmers, he scarcely roused them. They listened to his description of the arcane processes by which Parliament works and of the methods they could adopt to address those processes, but the very way in which he talked made it all seem unapproachably foreign, as if this were not something an ordinary farmer could have anything to do with. At one moment he said, 'There are lies, damned lies and developers' promises.' The room shifted at that, but its scarcely articulated anger did not emerge.

It then turned out that sitting there, in the front row but at one side, was the developers' own representative, Martin Robeson, Chartered Surveyor, of Littman and Robeson, acting on behalf of Carter Commercial Development. In the strangely theatrical way of these things, he looked exactly as a developers' agent should, with the sort of suede and black fur car-coat a developer in a TV drama would wear, a fairish goatee beard and a soft cow-lick of hair across his forehead. Questions began to drift away from the platform and towards him. He smoothly answered anything that was thrown at him, about the viability of the market and the ultimate purpose of his involvement here. 'Just because we intend to suspend the obligation to have a market here,' he said, 'does not mean to say that we will close the market.' There were jeers at that. A farmer from Punnett's Town – I could see his fields across the valley from Perch Hill – said, 'As I see it, big money and vested interests are overriding the interests of local people and that is a disgrace.' There was applause for that and Robeson took on the role of bogeyman for the meeting, so that whenever he stood up to speak a grumble of resentment accompanied his words.

I spoke to him afterwards. He was in a hurry, he had to get back to Oxfordshire, but he smiled and said they had received 'a lot of quite expensive advice about this, so we feel pretty confident we've got a good case. On a long-term trend, the market isn't viable. The graph goes down.' He angled his hand towards the floor. 'We can prove that. And if the market is not viable, then something else has got to happen to that land. OK?'

As we went out to the car park, I spoke to Buster Davis, a sheep farmer from Great Worge just up the lane. 'What do you think, Buster?' I said. He smiled. 'They've got all the money, haven't they? That's the problem. How do you fight that kind of thing? How do you fight it?'

As it happened, there was a stay of execution for Hailsham Market. The whole of the local community gathered against the predators: the council, the NFU, 235 farmers, scores of butchers, market traders, shopkeepers, peers of the realm, vicars. Together they orchestrated resistance to the supermarket and in the summer of 1997 the Bill to allow the destruction of the market, which had been moved by the developers in the House of Lords, was thrown out. Nothing was more effective than the speech by John Wrenbury, the non-stipendiary minister at Dallington, a bagpipe-playing Old Etonian ex-solicitor and hereditary peer. 'A market is a friendly place,' he said.

> You meet your friends there. You chew the cud. You have a drink – at least you do if the market is in Hailsham. A drink is important. You enjoy bangers and mash in the refreshment booths. You catch up on prices and are able to see what you are buying and what you could get if you were thinking of selling. Local gossip is exchanged. Old friends meet one another. Problems and complaints are aired. Know-how is exchanged. Crop performance and cattle prices are discussed.
>
> And that is only half the story. The wife is happy. She is taken into town to do her shopping. The local traders are delighted to see her and she meets her friends. Indeed, for some wives it is the only form of outing they get. One farmer wrote to me saying that being a farmer these days is a very lonely occupation because he has had to dispense with his farm hands. There is a great deal more to a market than an accountant ever realizes. It throbs with life and interest. Yet these are the very things that the Bill will do away with.

Lord Wrenbury and his friends won, a triumph for East Sussex. As Guildford market closed in 2000 and as Hampshire lost all of its markets, Hailsham became the only stockmarket between Ashford in Kent and Salisbury in Wiltshire or Thame in Oxfordshire.

Even then, safety was not assured. In 2006, another supermarket, Aldi, bought the land on which the market was held. Two years later, the multi-national company which had bought the market company itself collapsed. Only in January 2009 did a consortium of local farmers, the Hailsham Market Action Group, manage to buy the market company from the administrators. Now something strange started to happen. According to Janet Dann, secretary of the Action Group, the market which had been preserved as an important part of Old Sussex, the place where the small farmers of the Weald, the Marsh and the Downs could all meet and chat, started to grow and burgeon. It was not, on the whole, because large-scale farmers were starting to use it again but because the new type of farmer – small-scale, the urban escapee, with a few head of cattle and a small flock of sheep – found it to be exactly what they wanted and needed, just as their predecessers had done since the thirteenth century when this market began. Miraculously, it seemed that this social organism had just, by a whisker, been saved from destruction. As I write this in the spring of 2010, they are not yet out of the woods, however. The lease on the market site will run out in two years' time and if it is to survive the market will have to move, probably to the outskirts of Hailsham. In a recession, money is short and grants are becoming thin on the ground. No one yet feels secure.

That was one small stand against the flood. Otherwise, the local integrity of local places, sustaining local ways of doing things and local economic networks, out of which grow local social habits and a sense of locality in its richest forms – all that was draining away in front of my eyes like the suds in a soapy basin from which the plug has been pulled.

The farm sales were the worst. They were all, in their different ways, the same. Prinkle Farm, over at Dallington, went one sunny day. The tenant farmers had reached retiring age and their children had no wish to take it on. The 90 acres of heavy, steep clayland, even if run together in one unit with the neighbouring Carrick's Farm on the hill above it, could scarcely support a viable business. Closure was the most rational outcome: the tenants would go, the farmhouse would be sold or let to some ex-urban incomer, who could afford the rent or sale price, and the land would be absorbed in some larger enterprise. That was what usually happened. There had been a steady stream of farm sales around us. At each one, the life and belongings of the old farmers were sold away and the new money came in, nostalgic and acquisitive in equal measure.

At the sale itself, the crowds expressed nothing but delight as they picked over the remains of an existence whose time was now up. Hay knives, brass scales, horse harrows, implements whose use few could now recognize, cast-iron pig troughs, even cast-iron foot protectors, once strapped to their boots by men digging potatoes for too many hours at a stretch, had been dragged out of the back ends of sheds and half-abandoned barns and paraded here as 'mantelpiece stuff', 'bygones', 'collectables'. That was what got people excited and there was something vulturine about the excitement, a ravening for the carrion.

It was on a beautiful morning. The sun shone on the thousands of acres of Dallington Forest, arrayed before us around the flanks of the Dudwell valley. From the one or two farms scattered among the trees you could hear in the distance the heifers bellowing.

In the banky field opposite the farmhouse, three or four hundred cars and Land Rovers, some with trailers, were parked in ragged, shiny rows. And in the other banky field, the far side of the house, those innards of the farm were on show, long ribbons of intestine streaked across the field. All week the identical twin brothers, John and Peter Keeley, now sixty-four (Peter 10 minutes the elder), who had farmed here for the last thirty-seven years, had laid out everything they could find. The auctioneers had brought other lots in from other farms. Over 1,000 people had come to the sale, some from as far as the eastern end of Kent.

The chubby auctioneer, in shirt and tie, began. 'Look at that view,' he said. 'That's worth £50.' There was a buzz around him, a swarming so thick that only if you pushed in could you see him conducting the sale, lot by lot. These redundant objects were stripped of their dignity by the appetite for them. A group of three rather corroded old spring balances, used I suppose to weigh sacks of grain, lay on the grass in the centre of the clustering swarm. The usual rigmarole. 'What will you give me? Let's have a hundred. Eighty. Give me fifty. Fifty. I have fifty. Fifty-five. Sixty.' And so it went up. These things, neglected for twenty, thirty years, who could say how long, sold to someone with some kind of farm museum for £80. A hay-knife went for £40. Nostalgic money, or at least money that intended to make money out of nostalgic money – there were thought to be at least 15 dealers here – poured out into the sunshine.

At the end of the row, we came to something else. It was a large Massey-Ferguson seed drill, in good shape, capable of sowing 15 rows of seed at a time, a relatively hitech tool even if not of the most modern air-pressurized kind. It was a so-called combine drill which can inject nitrogen-based artificial fertilizer into the soil at the same time as the seed. It had obviously been kept in a shed and there was no rust on it.

'A fine Massey combine drill,' the auctioneer said. 'What will we have? Three hundred? Two? A hundred pounds? Fifty? Any bids? Any bids at all?' The crowd was silent around him. Some of them were clutching the bygones they had bought already, the cracked buckets, the unidentifiable implements, the blades from a rootcutter, the wormy hay rakes, the paring spade and the trenching gouge. Nobody wanted a tractor-drawn combine drill. Somebody offered a fiver and there was a little flurry, £7, £10, £12, £15, and there it stuck. £15 for a £3,000 seed drill of some real use if you were a cereal farmer but rather too big, at 12 feet by 8, for a mantelpiece.

'They would have got more if they had taken the wheels off and sold them separately,' a farmer said to me. But why did no one want it? 'Because you'd have to disc-harrow the field before you could use it. No one can be bothered with that.' Some tractors went for a couple of thousand and some mowers for a few hundred. The ewes averaged less than £40 a piece and the cattle under £300 a head, less than they had been bought for in the spring. It was all upside down: the decrepit was treasured, the useful and the healthy scarcely required.

I went to see the Keeley twins the next day. We sat and talked in their garden. They had never been apart for sixty-four years. Even when they did their national service, they had been together as radar operators in Fighter Command. 'He'll come on to say something,' John said, 'and I'll come out with it,' Peter said. They both sat on the edges of their chairs, both picking with their fingers at the other hand.

They were disappointed with the livestock but pleased at how the dead stock had gone. 'We were amazed at some of those prices,' Peter said. 'We couldn't believe some of it,' John said.

'A pig trow, cracked, for £50 . . .'

'Those horse harrows . . .'

'. . . we didn't even know we had half of it.'

'Those were the harrows Father used.'

'During the war . . .'

'. . . three hitched together.'

'The best one went for £25, was it?'

'£30, wasn't it?' They looked only at me throughout the duet.

This was the end of something. They had never sprayed their pastures 'because Father says it kills all the vetches . . .'

'. . . all the herbs.'

'We just run the mower over it.'

What Father says remains in the present tense. The sale had been 'a terrible day, a terrible emotional day'.

'My stomach was churning over all day . . .'

'. . . turning over and over.'

'A terrible emotional day.'

John was staying in a bungalow in Dallington. Peter was moving to a house in Ninfield because he couldn't afford one in the 'village on the hill', as he described the place he had always lived. 'That's what I'm most sorry about,' he said, 'leaving the district. It won't be the same.' Ninfield was just over six miles away. Something had come to an end.

The catastrophe was going on all round us, the ebbing of a tide, leaving a new and denuded geography. The Weald is a place of small farms, 100 acres on average, and more woods than anywhere else in England; of low incomes, intractable soils poor in trace elements and great beauty. The exigencies of the modern market have made this whole way of life virtually untenable now. The impact here falls not on farm-workers but on the small independent tenant farmers who have always been the backbone of the place. This is not a world of large farms with many hired hands. Each success in the past was a family success; each failure now, the failure of a family enterprise. It was always dairy country, but the Weald dairy herd had dropped by 40 per cent – 20,000 cows – in a decade. The number of full-time farm-workers had declined by a quarter in the same period.

The District Council made a study of four Wealden parishes which found that, of seventy-five farms in those parishes, not a single one could make enough profit to provide a market return on the land, labour and capital that were needed to run it. Low incomes meant no reinvestment, which meant lower incomes in the future, which meant no investment, which meant decline and collapse. Wealden farmers were already deeply in debt, far more than the national farming average, and there was already a danger that they would not be able to keep up with their interest payments. By the late 1990s many were hanging on, hoping for something better, not wanting to give up the lives and the places that made them what they were. The general farming decline compounded a situation already taut and tense with strain. Failure was being staved off week by week.

Yet these beautiful and unprofitable farms were worth a fortune. A pretty, comfortable farmhouse and 100 acres or so of wood and pasture were selling to the urban rich for anything up to £750,000. 'Why not let the market take its course?' a particularly cold-brained city analyst said to me one day at a drinks party where we were leaning against a fireplace together. 'Let the uneconomic farmers go out of business and realize their one real asset – the place. Send them off to live on the investment income and allow the commodity brokers and merchant bankers the pleasure of owning their place in the country. The real value of the Weald is now effectively as extensive suburbia. Any other outcome would be artificial.'

That was, effectively, what had happened to upstate New York or Connecticut. The farms had been abandoned, driven out of business by the massive cheapness of the prairie, and the trees had reinvaded the pastures. You could drive for miles through derelict landscapes where the old field walls net the scrubby woodland like the ghosts of a life once lived; where clapboarded family residences, occupied by Martha Stewart clones, preside over their own private, wood-hemmed clearings. Why not let the Weald become Connecticut?

I felt a big raging NO come up inside me. Why not? Because the essential quality of the place comes from the sort of farmers and woodmen who made it. Lose them and the place would be lost, the people would be lost, the life of the meadows and the coppiced woods would be lost.

A wrinkling realist sneer crossed the lips of the city man. 'What are you going to do about it, then?' I made him a speech. The place derived its qualities from being a labour-intensive landscape. Modern farming orthodoxy had seen labour-intensiveness as the great enemy, to be eradicated at all costs. But labour-intensiveness, as some people were now coming to realize, was the essential ingredient of the good landscape. It employed more people; it attended to the needs of the landscape more closely than a system whose priorities were set by the requirements of the machinery and chemicals used to run it; it generated other jobs; and it created places that other people would want to come to, see and stay in.

Didn't this provide a clear model for the future of any agricultural subsidy? Don't pay farms to sack people and then ruin the landscape with the giant labour-saving machines the subsidies allow them to afford. Pay farmers to employ people and to look after the more close-knit kind of landscape that people love. Hedges need hedgers. Payments should be made not per acre, as they then were, but per man employed. If there was a stimulated demand for farm labour, then farm wages would rise, the rural economy would benefit and so would rural services. Instead of money pouring into the pockets of machinery manufacturers and chemical conglomerates, it would be spent in the village shop. People made places good.

Of course, one can be sentimental about the local, about the old ways of doing things, and all too easily forget what was wrong with it. One day that year in Burwash, someone put up a large placard on his house wall. It said simply: 'Keep Burwash Clean and White.' Its author wouldn't talk to me about it. 'It was only up there 12 hours,' he said. 'To tell you the truth I didn't know what I was saying and I got so much strife about it. The village, the police, the lot, they were all giving me grief . . . So I'm sorry but it's no comment. That's all I've got to say.'

There had been a plan to set up an Indian takeaway in an empty shop – it was, by chance, the building in which my neighbour Shirley Ellman had once had her accountancy business – down at one end of the High Street. It had been met with a general burbling discontent, at least from those who lived at that end of the village. Only the author of the sign had come out quite so blankly with the racial purity line. The other arguments had focused on the traffic hazard of having a takeaway on that difficult corner, the lack of car parking and the possibility of litter. They pointed out that an application for a fish-and-chip shop a couple of years previously had met exactly the same objections and that one of the pubs had had a huge success with a Thai evening not long before, which was more than welcome. So the problem with the Indians was nothing to do with their race or colour. When you mentioned the poster, there was a slight, smiling turning away, 'disgraceful', not the right way to go about this at all. Only at the margins did the objectors move off on to hazier ground. They began with the idea that a place should be able to decide what it wants for itself. It was all about 'keeping Burwash a village, with a village atmosphere'. After a little downwind drift, this argument started to talk about 'a nice village' and then perhaps 'a nice English village, with a traditional English atmosphere' and then, if you began to ask what it was that might erode that atmosphere, the smell of the rejected alien began to waft across the Sussex fields.

Once I was attuned to the presence of this thought-chute in the ways of rural England, I started to detect it everywhere. By chance, after seeing the white supremacist poster in Burwash, I happened to be talking on the phone to a local lady. We were discussing the most insistent pressures on the rural landscape, roads perhaps, housing certainly, when without any warning she took a swerve into dangerous territory: 'Let's be honest about it, Adam,' she said confidingly, 'we don't want fuzzy-wuzzies and nignogs, do we, and you're bound to get them unless you're careful.'

Perhaps this is everywhere, virtually unstated, or stated only in private, the hidden but governing motivation of rural England. It brought back to mind a charming and wonderful man I knew once in Somerset. He was an eel fisherman. He knew everything about those secret and magical creatures, their lying-up places

in the rhynes and ditches of the Levels, the way to catch the
elvers as they came in off the Atlantic, smelling their way to
the sweet fresh water of the moorlands, and the excitement
of the silver eel harvest when, on a stormy and moonlit night
late in the year, with a big spring tide licking its way deep inland,
the adult eels, their backs silvered in maturity, make a sudden
rush for the ocean, in places teeming across the fields like a
disease or an infestation, rustling in the night past your feet
in the grasses. Everything about this man and his intimacy with
the place where he lived was to be admired. But then I asked
him what he did with the eels that he caught. 'I send them up
to London,' he said. 'We put them in boxes with holes in so they
can breathe, and when they get there it's the Jews that eat them.
We don't eat them. It's only the Jews in London that eat that
sort of thing.' No description of anything could have seemed more
alien to him than 'the Jews in London'. The phrase was shorthand
for everything that was not his and not known, not down there
with him in the wet, private world of the Somerset Levels.

I wonder now, in the light of this, if my own liberal attitude
is not in something of a muddle. There is little I value more than
the kind of local distinctiveness which oozed from the eel-man's
pores. He was the human version of the place he lived in. He
drew his sustenance from the almost purely local and as a result
had a kind of integrity which had its roots in the closeness with
which he was moulded to the place. But it was precisely those
same qualities which led him to think about the Jews in London
in the way he did.

Can you have one without the other? Does the fluidity and
acceptance of the strange, which is central to a liberal view of
the world, always eat away at the sense of the local which is so
valuable a part of the rural landscape? Is local distinctiveness
necessarily intolerant of the foreign?

Part Four

GROWING

Divorcing from the Past

THE INTENSE colours with which Sarah has planted her garden, a kind of beautiful smoky richness that sharpens in places into the brazen and the garish, have, I hear, drawn adverse comments. The garden doesn't go with the landscape. Its colours are much too strong. It's nothing like neat enough. It's untrammelled, unconfined. No one has ever done such a thing before and that is a good enough reason why no one ever should.

One summer at Perch Hill, we had a party and it felt like a prefiguring, in an almost ritual, ceremonial way, of the future. We had the party in Great Flemings, overlooking the valley of the River Dudwell, with the stepped and wooded ridges of the Weald folding back one after another to the north and east. The field had been mown for hay and the new grass was thick with clover like a flowery lawn. One open-sided tent for supper was at the top of the hill, another smaller one for dancing halfway down it. Simon Barden, who at that time drove the tractor here, had made a giant wigwam of a bonfire twelve feet high, from a pile of well-seasoned oak, cut two years previously from a tree that had begun dropping its boughs in the drought, just at the time that Stephen Wrenn had died. A few splashes of diesel in the foot of it and the thing burnt like a torch for six long hours, still there when I last saw it at three o'clock in the morning, a hot glowing ring like a fire-disc in the grass.

About fifty people came, and we drank and drank and danced and danced, untrammelled, unconfined, and the place itself, so soft and inviting in the last greying light of the evening, played its part. Deep in the night, long after any of us remembered what time it was or should be, the fireworks from the National Trust's annual party down at Bateman's in the valley below began to pop and glow in the sky beside us. Our hillside is about 250 feet above the valley floor and the rockets rose to meet us. Their burning trails were left behind them like the stalks of one gorgeous fire-flower after another, a giant night-garden flowering for our benefit. We had planned it like this; we knew it would happen; but this vision of fireworks swimming up towards us seemed then like a purely spontaneous eruption of beauty. We lay on the grass and stared at them, drinking in the booms they made, the old woods and fields reverberating with this sudden, man-made delight, as if the whole lit valley were saying, 'This is what life can be like, this is what the world can give you, this is how happy you can you be.'

That is a recognition that can come and go, be forgotten and remembered again, but on that party night the brightness seemed to be there for a while, irradiating us all, a ratchet clicked up, a point of optimism from which the whole future could be bathed in light.

By the time the decade and the millennium were coming to an end, Perch Hill and my life with Sarah had begun to have the effect I had longed for. I had become stronger again. I had regained my confidence. I could turn to other things. After writing the book about the restoration of Windsor Castle, I wrote a history of the ill-fated Millennium Dome, and then, as an escape from the poison of metropolitan politics, I spent much of a year away in the Hebrides, writing *Sea Room,* a book about the small islands my father had given me when I was twenty-one. That was followed by others which led me down different and intriguing paths: a book about Jacobean England and the

making of the King James Bible; a long sailing voyage up the
west coast of the British Isles, exploring the small offshore islands
on that coast and ending in the Faeroes; another about the
officers in Nelson's fleet at Trafalgar; an exploration of
Arcadianism in Renaissance England; and then, on the radio, a
series on the land- and seascapes of the *Iliad* and the *Odyssey*.

Looking back on this now, I can see it as a weaning off the
nurturing bosom of Perch Hill. I had slowly grown to the point
where I no longer needed it. This deep slow shift in our relationship to the place had taken seven or eight years. And as we
were changing, Sussex was changing around us. The idea that
we might fold ourselves into this pre-existent world faded away
as that world itself started to fade away. In its place, of necessity,
came something else: the idea that we could make here not
something that mimicked the past but something which could
take the best of the past and somehow fold it on into the future.

The result was that what we were doing at Perch Hill and
what was happening around us started to diverge. Our little
micro-world began to burgeon – with all the difficulties that
can accompany growth – as the larger world of the Weald seemed
increasingly to be forgetting its past, an implosion that smelled
of defeat and failure.

From time to time, that shift became fiercely apparent. One
Monday evening we had a meeting in the village hall about the
state of our lane. It had become An Issue. It was in a shockingly
bad condition; something needed To Be Done. The parish councillors of Brightling were there, about eight of them, sitting
around the central table. One or two parish councillors from
Burwash, which the lane runs into, were there too, as observers
and commentators.

Neon lights overhead. The hall is painted a slightly tangerine
version of magnolia. Framed photographs on the wall: cricket
teams from the 1920s, groups of gamekeepers and bellringers
from the decade before, the SS *Brightling*, a helicopter shot of

a large merchant ship, labouring in heavy seas. Some residents of the lane, a district councillor and Mr Furness from the County Council Highways Department sit in the outer ring of red stacking chairs. One of the village policemen is here, but not in uniform. There's hardly a tie or jacket to be seen. It is a cold night but the room is warm, if stark. This is, apparently, the best-attended meeting for years. The atmosphere is slightly lowering.

Although there is other business – planning applications made, planning permissions given, all the usual nitty-gritty – the reason we are all here is the lane and so the chairman brings that to the head of the agenda. The clerk reads out a letter written to the parish council late last year by Mr Furness, the Highway Manager. In it, he describes the root of his problems: money. The funds are no longer there to maintain rural lanes in East Sussex. As part of a £5 million cut the year before, lane maintenance was reduced quite literally to zero. No money, no maintenance. It was as simple as that.

The people in the room don't, on the whole, like the sound of what they are hearing. A larger symbolism and a fat wedge of significance seem to lie behind this relatively minor point. The lane – and by extension the villages themselves – are being deprived of funds. Is this the beginning of a long, slow strangulation by municipal authorities, which are essentially urban and have lost touch with rural needs and the rural frame of mind?

I'm sitting in the back row feeling guilty. The previous year I canvassed the residents of the lane about its future. I had heard about the budget crisis. I knew that various pressure groups, and even the Department of Transport, were engaged in new thinking about the future of small rural roads. Should the car, it was being asked, continue to have primacy in small lanes, at the expense of people living there, their dogs and children, and anyone who might want to walk or ride or bicycle in the lanes and generally enjoy their quietness and beauty? Why should car drivers be able to insist on universal access, dominating lanes even when they were not there simply by the threat of their presence? Shouldn't the lanes be a place for multiple use? That was the sort of prejudicial question I put to the residents and, not surprisingly, most of them thought, with one or two quite fierce exceptions, that closing the lane to cars at some mid-point was a good idea – as long as it didn't mean they would have to pay for it themselves.

That had been the residents' response. It was not the tone in the village hall that evening. There was a stronger charge to the atmosphere than you might have expected. After Mr Furness had spoken briefly, amplifying his letter, emphasizing that there were no plans to close the lane nor for the council to relinquish the principle of paying for its upkeep when it has the money, first the Brightling and then the Burwash councillors aired their anxieties. Could money not be saved on verge cutting and put towards road maintenance? Didn't the council realize that the lane was essential for people coming through from Dallington to the station at Stonegate? The alternative route was at least 2 miles longer. (Mr Furness estimated that to go round by the longer way, were the lane to be closed, would take an extra one and a half minutes.) One of the Burwash men spoke of 'wanting to return us to the age of the horse and cart'. The people of Burwash and Brightling did not want to be returned to the age of the horse and cart. There was talk of weight restriction and width restriction. One councillor had received a letter from a lady who suggested that signs should be put at the top and bottom of the lane saying that neither lorries nor burglars were welcome down it. That got a laugh.

Mr Furness gracefully accepted the points made and returned to his nub. There were no plans to close the lane but no money meant no maintenance. He accepted that in places it was rough. It was, in fact, he admitted, the roughest stretch of public road in the whole of East Sussex, particularly under the wood, where the trees sheltered it and the water lay, the frost penetrated and the surface fell apart. Mr Furness was managing decline.

It was my turn. As I began, I saw a look of long-jawed scepticism descend on the faces of the parish councillors. Oh, please, no, their faces said, not him. One held his head in hand, shaking it slowly. But I was not to be put off. The very roughness of the road, I said, was just what was wanted. It was, I went on, traffic-calming for free. The roughness meant people went slowly and that lorries hardly came down it at all. People could still use it if they really needed to but they couldn't race through it. Bollocks, the faces said.

I read out Kipling's description of this very lane in *Puck of Pook's Hill*: 'Puck jerked his head westward, where the valley narrows between wooded hills and steep hop-fields. A road led down there through thick, thick oak-forest, with deer in it, from the Beacon on the top of the hill – a shocking bad road it was . . .' That was its historical condition. Brightling and Burwash would actually gain something if it kept a lane in that condition. Think how their families would enjoy it. Think how lovely it would be to walk down in May, with all the flowers on the laneside banks.

None of it cut any ice. I was from a different world. The very instinct which had brought us here in the first place set us apart from the people who already were. Anyone who had lived in Brightling all their lives had no interest in rough lanes, or at least in the price to their convenience which rough lanes would bring. To have taken up that idea would have been to reverse the thinking of a lifetime, a frame of mind which focused on improvement, comfort and ease, not a thickly embedded sense of 'here'. We were after different things.

Sarah's and my presence at Perch Hill was yet itself a signal of something having gone. The ways of the past were draining away as we watched. For all our love of cattle and wild flowers, for all our attachment to hedges and woods and grazing made thistle-free again, we were the agents of change. We came from the world which disliked hunting, which made its living from books, newspapers and television, which had a hunger for the rooted and the rural because those things were not natural to us. We were the otherness we dreaded, tourists putting down roots, the toe of the eroding glacier, loving the country we were wearing away.

As it turned out, I lost the argument about the lane. Neighbours of ours driving low-slung Alfa-Romeos didn't like the rough-ness of the lane. They persuaded the council to resurface it. A team came to do the job one day, intent on laying a smooth mat of black tarmac on the old rutted and raddled surface. A man with a chainsaw went ahead of them, to lop off the limbs of the oak trees hanging over the lane so that the tarmac-laying lorry could fit underneath. Sarah and I found him as he was about to cut off the long, near-horizontal branch belonging to the oak tree at the corner of our garden, as if a stretch of tarmac could ever be worth twenty feet of oak. We managed to save that but nothing else the length of the lane. Its sides were shorn, the loppings chipped, the surface smoothed. In the month after the new tarmac was laid, there were three accidents in the lane, entirely the result of drivers feeling they could drive faster than before. We could sit in our kitchen and listen to the screech and crunch as the cars collided from the longest and smoothest of slides.

Things were changing here. Our own relationship to Perch Hill was changing. The early tight circle around our lives, the intense and sustaining need for it as an enclosure, which had driven us so fiercely in the early days, was becoming dilute. The urgency of our appetite for it had given way to something more integrated. The cord had loosened and there were sad departures. That spring Fred and Margaret Groombridge decided to retire and from the day they didn't come I missed them. To our great sadness, Will Clark died and was buried in Burwash churchyard.

At his funeral, the sun shone and the traffic roared down the village high street. From his grave you could just make out, a little more than a mile away, the last of our fields, the Way Field, dipping down between the woods towards the river. In the afternoon sunshine, the field was like a green flag, glossy with new growth, and a beautiful colour like the polished skin of an apple. It was neither browned off like those around it that had been part-abandoned, nor the chemical blue-green of those of some of our neighbours, who were pushing their land hard with nitrogen top-dressings. The colour of our grass was both richer and more natural than either, and for that lovely cared-for green I had only Will to thank. He had spent hour after hour on the tractor, topping the fields, nailing the weeds, bringing the farm back to health. He had been ill a long time. We all loved him.

Ken and Brenda decided to move away. He too had looked after us. A multi-talented man, he built beautiful walls and converted sheds to greenhouses. He could install electricity and was an expert joiner. He dug trenches and designed duck-houses. But more than any of that, he gave us all access to other things: a contact with the past and with Sussex; a sense of continuity amid all the changes we had imposed; any number of introductions to the local world of Groombridges, Keeleys and Bowleys, the underlying essence of this part of the Weald; and an unending sequence of jokes and stories rolling round and round the kitchen table.

Now Ken and Brenda were leaving for a house in Robertsbridge, travelling back down the road he had come up in 1942. The trust which owned their cottage was selling it. Ken had been at Perch Hill for fifty-six years, Brenda for thirty-nine. I looked out of the window one evening just before they left. Ken was walking there, Gemma his dog at his heels. Perhaps I was over-interpreting it but it looked as if his tread was heavier than before. He climbed the stile into his garden slowly and laboriously, conscious, I am sure, that this was the end of something. Gemma died the day they left and she was buried in the garden of their new house. Ken planted a rose Sarah and I had given them in the earth above her grave. People like the Weekeses wouldn't live at Perch Hill again. Ken felt that his moving was only the natural end of something that had been going on all around him for the last 10 or 20 years. 'I'm a stranger here now,' he said. 'It's all changed. There's hardly anyone I knew as a boy still here. It's sad really.' None of the five former dairy farms in our lane was now occupied by a farmer, at least in the strict sense of the word. We were all incomers now, deriving our livings from other places: the City, the media, estate agency. Only the Moodys in the nursery at the top of the lane, a lady who grew cyclamens and Sarah – to some extent – were actually getting a living from the land.

I found myself looking for the fragments of the past, for those elements in rural Sussex which had either gone or were on the point of going.

Late one winter I attended the annual South of England Hedge Laying Society Competition Day. Was this where I would find the past still alive? It was in full swing from about nine in the morning. About forty men – no women – from all over Surrey, Kent and both halves of Sussex, each with their 10-yard section of hedge, were confronting the task in front of them. The hedge was a tangle of overgrown hawthorn, spiky blackthorn, briar, bramble, one or two oaks, a goat willow here and there, old wire and mud. The hedge-layers had to cut, lay and bind this into order by 1.30 that afternoon. Then it was lunch and prizes in the Golden Cross Inn.

The grumbling discontent was already apparent when I arrived: the allocations of hedge had not been fair. As usual, the hedge-layers were divided into four classes: Novices, who had never won anything; Seniors, who had won at least a novice class; Veterans, which meant over sixty, but there was one man here who was eighty-six, swinging his chainsaw around his head as if it were a cheerleader's pompom; and Champions, who had at least won the Seniors before.

I had done a very small bit of hedge-laying myself and I could recognize easily enough if a hedge could be laid without problems. The whole task is to re-impose order on plants that want to break free of it. Hedge plants all want to become trees; laying a hedge is slapping them back into use. It used to be a common skill and it's not that difficult. The need for it disappeared when wire fencing arrived after the war: cheaper, quicker, easier and uglier. Before that, every farm used to have a man doing nothing but hedging and ditching all winter. Although it takes great skill to produce the beautiful pieces of craftsmanship the champions create, it is quite easy to make something that is crudely passable.

I quickly walked round the competition site. It was quite clear that the Champions class, seven or eight of the big men of the hedge-laying world, national champions among them, had been allocated by far the best stretch. They were cruising along, laying the plants – the pleachers – at a steady 30-degree angle to the horizontal, a pattern that was already looking like swept hair, combed out, regular.

Further down the ladder, things were not quite so happy. In fact, the further down you went, the unhappier they became. The Champions had been given the best stretch, the Seniors the next best – plenty of material, on the whole, but some big trees in amongst it – the Veterans worse than that, with thickets of briar on either side of the hedge, and the Novices, poor dears, stuck down on a hedge dividing the fields well away from the road, an absolute hell. The mood down their end was bad. Some of them were standing in cowy sludge halfway up their shins; others were faced with a stretch of 'hedge', only half of which was there.

Ian Johnson, a polite and dignified man, a retired town planner and novice hedger, slogging away at a few intractable hawthorns, had his hair plastered to his head with sweat. He looked suicidal. 'It is a bit of a . . .' he said, leaving that gap out of courtesy. Next to him, Fred Hoad from Dorking, 'a semi-retired venture

capitalist', was more forthcoming. Was he enjoying himself? 'No.'
What was his section like? 'The worst I have ever had and the
muddiest patch I have ever seen.' How would he describe it? 'A
bitch. Half of it's rotten at the bottom.'

I asked the president, John Wilson, why the best had been
given the best and the worst the worst. He put his finger in his
mouth like a boy found out. 'You couldn't trust the Novices with
the roadside. It's got to look good on the roadside.' That was
part of it, but why weren't the Champions given the Seniors'
stretch? No answer to that but we all knew what was going on:
unto every one that hath . . . the old story.

Up with the Champions, life couldn't have been sunnier. Des
Whittington from Blackboys in Sussex said, 'They call me Dick.
You can understand why.' 'Yes,' says Alan Ashby, one of the greats,
next to him, 'and that's the right word for Whittington.' Ashby
and Peter Tunks, from Horley near Gatwick, national champion
in twelve out of the last fourteen seasons, had just done two
miles of hedge-laying together on the M1. Where was that?
'Between Junctions 8 and 9, northbound. Another lot were doing
the southbound.' What were they like? 'Rustic,' Ashby said.

It was all like that: heavy joshing. At 1.30 the hooter went to
mark the end of the time. Half the men didn't pay attention
and carried on working for twenty minutes or more. The chief
steward couldn't be found – one theory was that he'd gone off
to lunch. Eventually we all ambled off to the pub. The hedge we
left behind looked wonderful, a neat, twiggy-basket line of living
plants, stockproof, resilient and with a future.

There was a stinging surprise in the tail. Peter Tunks, one
of the world's best, had produced, as ever, what looked like
an immaculate piece of work. He was not happy with it – 'a
bit gappy' – but he never is. After lunch, the prize-giving.
The Novices and Seniors were announced. No controversy.
John Wilson, the president, won the Veteran class. Everyone
was pleased. Then the Champions class. In reverse order: 4th

John Savings, from Oxfordshire, one of the Midland boys; 3rd Dave Truran, from Burgess Hill, who'd been struggling with his kit all morning – the blade fell off his chainsaw – but he'd come through; 2nd Alan Ashby, all smiles. At the table I was sitting at, everyone was predicting Tunks. 'It's got to be Tunksy, hasn't it?' 'And the winner is Robert Graham.' The young man from Five Oak Green near Tunbridge Wells received his trophy and thanked everyone blushingly. Tunksy was unplaced.

I started asking afterwards why the star had come nowhere. Suddenly, a man in the pub corridor, I didn't know who he was, flew off the handle. 'We don't want you coming here and questioning our judge's conclusions.' There was a distinct red patch in his neck. 'A judge's decision is final. What do you know about hedge-laying, anyway?' I found Tunks. He raised his eyebrows. 'It's all part of the game, isn't it,' he said. Mystery on mystery. But what, really, did any of this have to do with the rural past? Could it really have been like this before? Had even hedge-laying entered a modern, de-natured condition?

Of all the trashed inheritances around us, none was more shocking than the big wood that surrounds the whole eastern end of our farm. It was, in my eyes anyway, a wreck. It was a symbol of precisely what not to do if you are entrusted with a place that is full of ancient significance and layered meanings. Its modern condition is a monument to hubris, to the pitfalls of arrogance, to the careless substitution of what looks like a good scam for an apparently less productive and outdated way of doing things. It will take centuries to mend the damage that has been done, if it ever can be.

High Wood used to clothe the sides of the Dudwell valley. It formed a wooded belt between the last of our fields and those, a mile away, belonging to Bateman's. The land on which the trees grew above the fertile valley floor is poor, mostly acid, sandy heath, full of heathers and broom. It could never have been cultivated. The wood was certainly older than any records of it, a fragment of the forest that once covered the Weald from edge to edge, a place of deep stability.

As Kipling wrote in *Puck of Pook's Hill*, these were 'the woods that know everything and tell nothing'. They became for him a sort of reservoir of the English spirit which could emerge from the leafy shadows for an hour or two and then blend back into them, a place where the gates were down between the landscape, the idea of history and the feeling of other lives and other spirits inhabiting the world we call ours.

The wood had most of its big oaks taken out during the Second World War. In the early 1950s it was sold to the Forestry Commission, on a 999-year lease, a legal fiction allowing the previous owners to retain shooting rights. The Commission, in its words, then 'sprayed off the woody weeds', in other words killed the broadleaf trees of which the wood had been made for millennia, and planted most of it up with a variety of softwoods, spruces and a kind of tall dark pine called western hemlock. In one or two places it left the old chestnut coppice and in others, where it was too wet to make wholesale replanting either practicable or financially worthwhile, it left the rough old mixture of ash, field maple, hornbeam and thorn.

The Commission put in the new roads to extract their valuable resource – roads which always seem, incidentally, unnecessarily vast for their purpose. The old draw roads by which timber was taken out are always subtly sewn into these woods, not fearsomely driven through them. The Commission then thinned its crop from time to time and waited for the final clear cut to make its killing. It never came. The hurricane in 1987 flattened acre after acre. Large parts of the wood were virtually inaccessible for years.

High Wood has now been in public ownership for over fifty years and has yet to produce a single penny of profit.

The wind-devastated zone has been cleared, the ground has been scarified, new rabbit fencing has been installed, new seedlings of Scots pine and other species have been planted. In parts, the Commission has allowed natural regeneration to occur. Birch trees dominate on the sandy soil. There are one or two oaks. Here and there an older broadleaf tree has managed to survive the years of repression. Bramble entanglements fill the spaces between the school-age trees. Dead stumps of sprayed-off chestnuts poke through the thorny rubbish like the vertebrae of flensed whales.

But how much is gone! Everyone used to think of this as their wood, to walk in, make love in, make camps in, hide in, dream in, be in. But that soft communal assumption, an ancient understanding that goes beyond the law of property or contract, had been neglected and denied here by an agency said to be acting in the interests of us all.

In 2001, as a result largely of the centralization of modern farming and marketing practices, foot-and-mouth disease spread rapidly and catastrophically across the country. It never came to Sussex, but we were caught in the outer ripples of the disaster. The absurdity of what was happening struck me one week when I saw on television the epidemiologist Professor Roy Anderson, of Imperial College London, telling the House of Commons Food and Rural Affairs Select Committee that he had asked the civil servants in Defra, the Department for Farming, the Environment and Rural Affairs, for the database of farm locations by which the spread of foot-and-mouth disease had been monitored. They had sent him the data, but the co-ordinates they provided left him perplexed. Half the farms he tried to look up, as he told the committee, 'were out in the North Sea'.

This had the MPs in stitches; there were hoots of laughter at the idiocy of computers and/or civil servants, or both. But on the ground, it became more than a joke. Foot-and-mouth hadn't been anywhere near East Sussex, but we, like the rest of the country, continued to labour under the most pedantic licensing scheme any computer or bureaucrat could have devised. All movements of animals between one holding and another needed to be licensed and supervised by a vet, a bore for us, but also more than that.

We had some lambs in a field by the church at Netherfield, a couple of miles away, which needed to be moved to fresh grazing 300 yards down the road. The new grazing was nominally in a different holding. They needed a licence. We applied for a licence at the trading standards office in Lewes. The trading standards officer refused to grant it. Why? Because the field next to the church in Netherfield was 'in a dirty area'.

'A what?'

'An infected area. The computer says it is.'

'But there has been no foot-and-mouth in East Sussex.'

'We can't go against the computer.'

'So what can we do?'

'Talk to Defra about it.'

'We spoke to Defra. Nothing to do with them. We should speak to Adas, the privatized agricultural consultancy (motto: 'Helping farmers to help themselves'). Adas said it was nothing to do with them. We should speak to the trading standards people. We spoke to them again. They said we should speak to Defra again, to their map division. The map division required all the papers to do with the case. We weren't allowed to send the details by fax. Everything had to go by post.

It is one of the more frustrating aspects of Defra culture that its relationship to modern technology is chaotic. Not only were they relying for their decision-making on computers that confused East Sussex with Cumbria, and Oxfordshire with the Goodwin Sands; if their computers broke down, which they did at steady intervals, the officials were not allowed to turn to pen and paper until the machines had failed to work for more than eight hours.

There were offices all over the country where officials were sitting for hours at a time twiddling their thumbs. Nor could any movement licences be faxed out. They too all had to go in the post. And they couldn't be sent to the farmer who needed to move his animals. They could only go to the vet the farmer had nominated – yet another stiffening of the system. Individual vets had received forty movement licences in a single post, which they could not possibly deal with in a single day.

So what happened to the Netherfield lambs? Their grass ran out. In the two weeks the system took to work through its coils, two of the lambs died. The rest lost condition in a way no farmer likes to see in his stock. And, of course, there was not an ounce of compensation.

It was laughable. Sheer delay, the reports estimated, was responsible for the slaughter of about three million animals which would otherwise have been unaffected. There was talk of making the licensing scheme a permanent feature of farmers' lives, to be controlled by officials who know as much about the geography of rural Britain, and the lives of those who occupy it, as they do about the Matto Grosso or the hunting habits of the Nambikwara.

The foot-and-mouth disaster occurred against the background of the campaign to ban hunting. I have never been a hunter myself and have always been more interested in feeling alive in the world than in killing other inhabitants of it. But I have never been one with a fierce ideological objection to hunting or shooting. I once spent a beautiful day chasing foxes on foot with the Blencathra Foxhounds in the Lake District, and, as I described on page 89, the last animal I shot – or to be honest shot at – died of a heart attack.

I had missed the hare I had aimed at but it then died of fright. A shameful thing.

Twenty-seven years went by. I sold my guns. I even sold the guns my grandfather left me, a beautiful matched pair, whose value was poured into the overdraft. Shooting belonged to a previous existence. Every morning at the pond in our farmyard, when we went out to feed our ducks there with a handful or two of mixed corn, there was a small, although nowadays increasingly large, train of refugee pheasants that came silently and with great dignity after the squabbling, nattering ducks had barged their way in towards the scattered corn. There is a high-stepping gait to the pheasant which, especially after rain with their high-coloured feathers all bedraggled with the wet, turns them into White Russians, aristocrats fallen on thin times. The ducks are like fishwives next to them, English born and bred, indifferent to the higher forms of melancholy.

Of course I felt sympathy with the pheasants. They looked like portraits by Graham Sutherland, or members of an ancient family eking out an existence in a National Trust house which was no longer theirs, Cavaliers left over after the King's head had been cut off, their dignity somehow bound up with their sense of impotence. Why did they exist, except to provide a moment's diversion for a well-heeled, amateur killer, who was either thinking of what would happen at lunch or slightly hazy with its after-effects? At the Day of Judgement, when the pheasant-shooters are lined up on one sside of the court and the pheasants on the other, could there ever be any doubt which would be sent to heaven and which to hell?

And I'd been beating once. I had somehow imagined it would be a noisy, whooping business, but it wasn't. The other beaters and I swept in a curving line through a plantation in almost total silence. Only the sticks knocking against the tree trunks could be heard and the occasional high-pitched whistle for a dog. It was as though we were drawing a net in, ushering the pheasants into the foot of the bag. They scuttled ahead of us, silently and creepingly. It was a strange piece of theatre, three groups of us involved: the silent, knocking men, the silent, creeping birds, and the silent, waiting guns.

One or two birds shuffled into flight, blustering through the larches and then out into the field, gathering height and then over the wood in which the guns were standing. The air smacked with the sound of the cartridges going off – a big, flat-palmed slapping of the air – and the pheasants flew on, virtually every one of them, completely unaffected. The beaters were guiding hundreds of birds out of the wood over the line of guns; the men in suits and ties were banging away at the pheasants over-head and missing. The keeper and his team of beaters stood and watched as the carefully raised, fed, nurtured and now driven birds flew towards their destiny. Bang, bang, bang, bang, went the wood, like an upper-class laugh, the pheasants flew onwards and the keeper turned away smiling. 'Useless,' he grinned.

Then, in the winter of 2002 one of my friends and neighbours asked me if I would like to come for a day's shooting. 'Why not?' I said. So I turned up, Saturday morning, ten to nine, a beautiful day, diamond clear, frost on the lips. I was in jeans, a T-shirt and a fleece jacket. Everyone else was in tweed plus-fours, tweed jackets and waistcoats, collars and ties, long socks with garters from which little flags hung down below the turned-down stock-ings, tweed caps, fresh faces, lovely welcoming smiles. My hat was a blue nylon thing with the word 'Dolomite' embroidered on it. I felt utterly at home.

My neighbour Simon gave me a gun, a cartridge belt and a quick talk. Don't shoot anyone was the burden of his message. Safety, really, that's the key. Lovely. Simon exudes a sense of heavenly contentment and generosity. Everything was going to be marvellous.

On to the first drive. I was between Frank and Tarquin, lined up 80 yards apart across a grassy bank on which the frost still lay. In front of us the margins of a wood. We waited. Having not pulled a trigger for quarter of a century, I was a little nervous. The tock-tocking of the beaters' sticks against the trees. The first scatter of songbirds coming out of the wood high above us, flit-tering this way and that. I suppose in Italy we might have shot them. The guns stood silently. I put a pair of cartridges in the breech and waited. I slid the safety catch to and fro. I fingered the triggers. I tried a few practice swings: up to the shoulder, both eyes open, keep swinging through the bird, don't stiffen up.

We waited. I was staring over at the far corner of the wood where the songbirds had for some reason gathered. It was a dreamy morning. I wasn't concentrating. 'ADAM,' Frank shouted at me from below. I looked up. A hen pheasant was sailing calmly over my head. I watched it go, saw its breast feathers above me, raised no gun to it, listened to it churring away into the distance. Not a good start.

Gun now in two hands, left thumb aligned along the barrel, right finger on trigger guard. Oh Christ, here it comes. The cock pheasant, curving gently towards me out of the top of the wood, dark against the Eton-blue sky, a steady easy straight flight right into my line of fire. Up comes the gun, safety catch off, barrels over the bird, swinging through, squeeze, keep it going, PHLAM! Ringing in the ears, the pheasant flying on past me, calm itself, untroubled, perfect, a clear escape off towards the bobbled, tree-filled distances of the Weald and then again PHLAM! Frank, swinging easily from his stance below me, picks the pheasant out of the air, all its poise crumples, its head goes jerking back as if hit by a wall, its arranged body now a bundle of washing thudding into the earth. It sounds like a slightly deflated football.

All morning I fired and missed. I didn't mind. All the other guns gave me sweet advice. Think of it as if you are flinging a handful of gravel into the air. It's a chucking motion. Don't aim. Just throw it. None of it did any good. I stood and watched a woodcock fly past me, darting and quivering between the trees. A rather chubby fox trotted slowly through the line of guns. Any pheasant that came past I fired at to no effect. A beater standing next to me, watching this embarrassing performance, said quite courteously that I shouldn't poke my gun at it: you wouldn't get them with poking. He had been picking up for Prince Edward at Eridge the week before and the prince had apologized to him whenever he didn't kill a bird outright. So I felt perfectly happy apologizing for not killing any at all.

We walked back for a steaming lunch. A couple of glasses of delicious claret. Suffused warmth and contentment. This was the style for the new millennium: a return to the 1920s. Was it that sudden acceptance that made the difference? Or was it the claret? Who can say? But difference there was. I stood at my peg for the first drive after lunch. A steep bank of beech, holly and hornbeam rose in front of me. The fallen leaves were pulled about by the breeze. The now-familiar knocking of the beaters' sticks. A slight, early evening mist creeping between the branches, the snuffling about of their dogs. A hen pheasant comes over, steers to my left, I follow it round, swinging through, chuck the gravel, squeeze the trigger and the bird tumbles to the ground dead. That clenching, air-punching YES which is not, I think, quite the proper thing to do. And then another, a few minutes later. And then far too excited, shooting low in front of me and told off by the Guards officer standing to my left. 'Rather too low, I think,' he said in the most courteous of voices. 'Quite, quite,' I said, 'sorry, sorry,' rather surprised to hear my voice now sounding like that of a captain in the Grenadiers, but at the same time and quite suddenly realizing I hadn't had such fun since surfing at Polzeath the previous summer.

But shooting is not hunting and one weekend, as the prospect of the ban was looming over the countryside, I had myself hunted. The Coakham Bloodhounds were meeting at a farm on the shoulders of Ashdown Forest in the north of the county. It was a slightly drizzly morning, but all the usual hunt scenes were there: expensive-looking horse-boxes; men shuffling out of jeans and into riding breeches at the far side of Land Rovers; women putting lipstick on lips and hair in nets; people, in short, getting dressed for the hunt as if for an appearance in *Barry Lyndon*,

21st-century barristers, shipbrokers, farmers and builders suddenly emerging with their horses like characters from a Stubbs.

These were the people whose quarry I had volunteered to be. The bloodhound is a solemn-looking figure, the most droopingly melancholy of dogs, curtain-jowled and heavy-eyed, with feet the size of lions' paws and an appearance that suggests what he is: a body in pursuit of a nose. Bloodhounds are smelling machines, originally used to sniff out the blood of a wounded deer, extraordinarily able to pursue the most fugitive of scents across field, moor, wood and even rivers and well able to chase after the sweet healthy smell of men on the move.

There were four of us that Sunday: Adrian 'Blobby' or 'Bear' Paice, the man-mountain of a Quarry Captain; a super-fit, ex-Royal Engineers bomb-disposal expert, Derek 'Piers' Barnes, who from time to time would declare his love for 'Blobs' and kiss him on both cheeks; Robbie 'Freddy' Miles, who had once run a line for the hunt and then jumped on a horse and followed the hounds after his own scent; and me.

The mounted figures from the eighteenth century drank something alcoholic out of plastic cups, while we quarry stood waiting for the hounds. There was a slight tinge here of class drama: runner-squaddies headed for the front line, hunter General Staff inspecting them from horseback, and saying from time to time, 'Well done, thank you so much, thank you for all your efforts.' We Baldricks soon had ourselves nuzzled and licked by the pack, huge soft bloodhound noses sniffing every part of us, the hounds putting their paws up on our chests, noses discovering who we were.

Then, as the language of this wonderfully retrospective corner of Englishness expressed it, 'for the first scurry, the quarry are enlarged twenty minutes before the hounds are laid on'. It was excitement itself. We ran nine miles in front of the hounds, through woods, wading through rivers, up the steep-sided valleys that cut into Ashdown Forest, scrambling through hedgerows, jumping over tiger traps. Two or three times, at pre-arranged places, we stopped and allowed our pursuers to catch us. The music, as it is called, of a pack of hounds on your trail, faint at first, carried up the valley on the wind, gradually getting louder, the honking, baying joint passion of the pack coming after you, is, no doubt, a faintly disturbing thing: it is not difficult to imagine it as happening for real, the sound of imminent death.

It is soon followed by the noise of the hounds crashing through the wood in your wake, nothing to stop them, before they finally break cover, up the field to find you, lips and ears flapping at each bounding step, tongues flying like commodore's pennants, and then the arrival, the death by a hundred licks. 'Good boys,' we all say mindlessly. 'Gooood booooys,' handing out the dog biscuits we have been carrying in our pockets, lick, sniff, lick, wag, as the field arrives, massively out of breath, red in the face, sweat on the flanks of the horses. After a few minutes' rest, the runners set off again, twenty minutes' start on the hounds, the whole ritual re-enacted. It is – as it is meant to be – a strange picture of happiness.

I've rarely had such fun: get-fit, urban runners spend the day with go-for-it rural hunters; no one and nothing gets killed; something of the tradition of hunting a pack of hounds is kept alive; the management of the landscape for hunting continues; and everybody is happy. Was that right? Not entirely. It was so clearly a game, only one step removed from paint-balling. A runner obviously loves it but the hunters don't really like it. They told me that manhunting is too hard and too fast, with none of the hanging about, the chat, that characterizes a fox hunt and, anyway, as one of them said to me, 'Hunting a jogger is like kissing your sister: not the real thing.' The coming ban would mean that the real thing would be denied them and this would be all they had left.

In all these various ways – the smoothing of lanes, the death and departure of the old, the commercialization of rural crafts, the management of woods, the bureaucratization of farming, the banning of the hunt – in all of them, the future here was being divorced from the past. There was only one exception, a place where landscape and memory still joined up. I went one day on the village expedition to visit the graves of the Burwash men in the cemeteries of the Ypres salient. About 125 men from Burwash and Burwash Common were killed in France and Flanders during the First World War, 18 of them in the muddy slaughterhouse of the Salient, the British Verdun. In a village of about 2,500 people, this death toll represented about 10 per cent of the men, or perhaps 30 per cent of the relevant generation. It is mortality on a scale you find in French villages but rarely here and it is thought in Burwash, although no one is sure, that proportionately this village suffered as much as any in England.

Perhaps that is the reason some care is still taken. The war memorial in the village street is fitted with an electric light in a small lantern crowning it. The light is lit on the days of the year when a Burwash man was killed. I have always been brought up short by that small signal. You're driving back after a summer day, your mind is full of other things, the tape machine is on, the evening is fading from the sky, you come around the corner, past the Bell Inn and the church, and there at the side of road, that point of light, a day of death, a sudden individualization of mass catastrophe.

In the soft morning light at half past five in the village car park, the rain was solid, sluicing off the roof of the bus in which the fifty or so of us sat, perfect quagmire rain, the water flowing in tea-coloured rivulets down the lanes, clay in suspension, the Weald running away to its rivers.

This had been advertised as a pilgrimage, and in part, for sure, it was that, a return to some sort of essential point, a base to be touched and recognized for what it was, but like any pilgrimage, it was also an outing, £40 each, three-course lunch thrown in, plenty of wine, a chance for a bit of duty-free. All day long we moved in and out of those categories. Sir Frank Sanderson, a charming baronet in his mid-60s, who runs the Burwash branch of the British Legion and had arranged this trip, spoke to us from the front of the bus about the war and the casualties – quoting Sassoon, 'the sullen swamp', 'the sepulchre of crime' – and recommended making a beeline for the restaurant as soon as we got on to the ferry 'because otherwise you won't get a table'. And he thought that the ferry loos would be the best we would find all day.

Once over in France, the Sussex party didn't think much of the Flemish landscape – 'boring' – and couldn't see any evidence of the Flemish farmers being forced to put any of their land into set-aside but noticed that the farmhouses looked in suspiciously good nick. Everything rural England suspects about the treachery of Europe was being confirmed before their eyes. There were dark mutterings about Germany having been stopped twice but who was going to stop them this time? Sir Frank outlined the Schlieffen plan by which the Germans in 1914 intended to mop up France in a month and went on, seamlessly, to describe the Schengen Area in which other Europeans no longer have to show their passports. 'Some might say,' he went on, 'there is something frighteningly similar about those two names.' But this was something of a tease and Sanderson for one stands above Europhobia. Much later on I saw him in the chill beauty of the German cemetery at Langemarck, laying a bunch of poppies on the memorial there which he had picked the day before in a Burwash garden.

As we approached Ypres that morning, the tone had deepened. 'We are now,' Sanderson said, 'on the road our martyrs took. And as we go east, remember, we are going towards the guns.' He spoke in the present tense. There were hop gardens alongside the road and pollarded willows. Small herds of Friesians and Charolais were lying in the lush meadows. Little drainage ditches were full of flag irises and water lilies. 'The man we are visiting,'

Sanderson said, 'is Albert John Morris, a private in The Buffs who was wounded in June 1917 during an attack at a place called Spoil Bank or Buff Bank at the southern edge of the Salient and then brought back to a small hospital next to the railway here, where we now are, at Lijssenthoek, where there was a casualty clearing station.'

Morris was twenty four. His widowed mother, Mary, lived at Rock's Cottages in Burwash. Her son died of his wounds on the evening of 15 June 1917, at about 8.30. Sanderson read a letter from his commanding officer to Mary Morris. It was written in a terrible rush – 'There is a lot to do on the battlefield.' Clearly the captain knew little about the dead soldier. 'He was a good man and always willing to do his share,' Captain Morrell wrote but for any hard information he recommended she get in touch with '17 CCS', the Lijssenthoek Casualty Clearing Station.

Where there was once a hospital, there was now a graveyard. Wistaria was trained along its low brick walls. Giant cedars stretched their arms out across the row on row of white Portland headstones. The grass was mown like a putting green. One or two of us got out of the bus and found Morris's grave. The vicar read a prayer over it. 'Into thy hands, O most loving father, we commend the soul of Albert Morris, whom we remember here today, humbly beseeching Thee that he might be precious in thy sight.' Precious in thy sight. Sir Frank's wife, Lady Sanderson, planted a little sprig of rosemary, which she had brought with her from Burwash.

It all made for a tiny gesture in a Flemish field, surrounded by the ridged lines of potatoes and with a warm European wind blowing across us. Nothing in itself, or almost nothing, but as the day went by and, one after another in these cemeteries scattered across the Salient, the Burwash men were remembered and the little rosemaries planted, for 'Tinto' Park, for J.H. Sweatman, Stanley Taylor, George Jeffrey and Arthur Chittenden, one by one these tiny acts of remembrance became something else, a reconstitution of that village generation, binding them back together, re-establishing, for us and for this day anyway, the links between them.

Why was this so moving? Perhaps because these men, who had been brought up together, who had been at the village school together, who knew the same fields and woods and girls, who were from one clumped-together genetic stock, now lay so widely scattered across this anonymous European plain. Each time the bus stopped and we climbed out into another cold, immaculate graveyard, it was as if we were visiting another fragment of a time and a place that had been blown apart. Each Burwash body was like a single autumn leaf, cast away in this place that meant nothing to Brightling or Burwash or the people who lived there. I have never thought fields so foreign. And I felt for the coldness of this distant death, these Weekeses, Groombridges and Keeleys, so far from the places they knew, and so painfully alone.

The Very Opposite of Poisonous

AT HOME, though, things were not so entirely retrospective. 'Will you look at that?' Simon Bishop said to us all one evening. Farmer, cook, agricultural lecturer, father, husband, wit and Sussex man into the very pit of his boots, just turned forty, slightly thin on top, smiling a gee-shucks smile like Wallace and Gromit's, he took a wing rib from the oven and placed it on the table in front of us. It was from a pure-bred Sussex steer he had raised himself at Ivyland Farm near Netherfield. It was a Sunday evening. Sarah, I and our two daughters, Simon's wife, Tessa, their daughters Beckie and Holly and their son James, were all there around the table. The early winter night was dark outside. There were candles on the table. 'Quiet, everyone,' Simon said, a sort of joke priestliness on his face.

We looked reverently on the glistening rib, its glazed surfaces, its ruckled fat, the oozing flanks of meat. This was beef on the bone, months after such a thing had been banned amid the BSE crisis. We gazed on its heroic naturalness, its grand status as a hunk of nature. 'I feel I'm making love to it,' Simon said.

'Will you get on and cut the thing?' Tessa said. 'I'm starving.'

'Wait,' Simon said with a sudden inrush of authority. 'We must let it rest.'

'What do you mean, "rest"?' Tessa said. 'It's dead.' Simon, the priest-magician, easing and wooing with his spell-casting fingers, said that the juices had to go back into the meat. It was quite clear he didn't know what he was talking about. But we waited anyway as the smells wafted around the room. Cooked in an oven that was fuelled from logs cut on this farm, surrounded by leeks, carrots, potatoes, beetroots, red and white onions, marrow, tomatoes and parsnips that the Bishops had grown here and harvested that morning, this was a version of completeness. It was how things should be.

Nothing had come between us and the beef except the killing, the butchering and the roasting of it. Here was the man who had brought it up, who had tended it until it was ready, who had invited the butcher down to have a look, awaited his verdict, which was whispered, confidential ('It does you credit, Simon'), who had chosen the hauliers to take the animal without fuss on the four- or five-mile journey to the Brownings' abattoir at Broad Oak, who had it back here now to eat with his family and friends.

Perhaps only what was threatened needs celebration. The beef feast at Simon Bishop's farm on that Sunday was such a moment because so few could enjoy it. If this had been a restaurant and he were not the grower, he would have been prosecuted for that dinner and heavily fined. Any chance of BSE being present in his animals, let alone us contracting new-variant CJD from eating the beef, was literally nil. The wing rib came from Simon's single suckler herd, in which the animals, once they wean themselves from their mothers, are fed on nothing but the grass in the fields outside the window and the silage made on the farm. Nevertheless, the law applies to this beef as to any other. Beef on the bone was to be treated as poisonous by the authorities, but everything about this scene, and the Bishop family who had created it, was the very opposite of poisonous. Here was a man who believed that the life of a beef and sheep farmer in the Sussex Weald could still be happy and sustained.

Simon steered me towards some government grants and so, for the first time since we came here, I engaged with the dominant and dictating force in the modern landscape. Subsidies now shape the rural world, but the irony was this: to do the thing that was most like what farmers do today, to book into that money-for-nothing system, felt as alien as anything I had ever done to this place.

I hired a consultant, an expert in navigating the bureaucratic shoals with which government money is guarded. Together with him, for an enjoyable morning, we decided how to work the system to our advantage. It was a revelatory experience, a brief exposure to the frame of mind which farmers must now enter if they are to survive in the radically distorted agricultural market. Civil servants would decide if I was to get the money I wanted and so, like other farmers, I had to second-guess their intentions and even prejudices. Would they really contribute to the cost of a new barn roof? Could we say a new barn roof would perform some educational purpose? What about removing that horrible old fence? Could we say it was an eyesore visible from a public path? I gathered, from my consultant, that certain areas had been or were likely to be 'under-applied' that year and so I shouldn't hold back from going for as much as possible. I was to exercise canniness, creating an attractive package which the civil servants would 'buy'. It was not, in terms of the money that some East Anglian barley barons were then receiving from the Exchequer, or even what your average farmer was getting, a large amount. But I certainly wasn't sniffing at it.

We put seven out of the eleven fields into a scheme run by the Ministry called Countryside Stewardship. We agreed, in essence, not to overstock the pastures, nor to fertilize them so that the weaker plants would have a chance and, if the fields were to be shut up for hay, to cut them very late so that the flowers could set seed.

For this, which was basically agreeing not to push the land hard, the Ministry would pay me £70 a hectare, with a supplement of £30 a hectare if the fields were small and another supplement of £40 a hectare if there was a serious thistle problem, which there was. We would also agree to have visits from schools on six mornings a year to show them what was going on and for that we would get a payment of £400 a year. The result was that we would get a total subsidy on just over 16 hectares, or about 40 acres, of a little over £2,000 a year over the ten years for which we entered the agreement. On top of that, the government would pay just about half the costs of reinstating and repairing ancient hedgerows and gateways to the tune of about £5,300 over the following five years. In all, then, I would get from the taxpayer over the next decade about £25,000.

What did we think about that? The first thing was that these figures were laughably low compared with other places. There were seven English farmers who received over £500,000 the previous year alone and you could be pretty sure that bung would have come on top of almost equivalent profits on grain sales. If, now, instead of putting my 16 hectares in the Countryside Stewardship scheme, I had decided to sow linseed I would have received £650 a hectare a year in subsidy, which would add up to more than £100,000 if I did it for ten years, four times the amount my flower-conscious management of grassland might get me. If this were my profession, which do you think I might go for? £2,000 a year for wild flowers or £10,000 a year, from the same area of ground, for linseed oil? No one should be surprised that farmers with flowery meadows were ploughing them up, and would be ploughing them up the following winter, to plant the miracle subsidy crops. They were being paid to do so.

The national picture was totally lopsided. In the whole country that year the government was putting £100 million of its receipts from the Common Agricultural Policy towards improving the agricultural environment. That sounded quite good, but it sounds less good when you hear that they were spending £4,300 million on agricultural production without any thought of the environment whatsoever. So it was ridiculous of me to be feeling any qualms about accepting some of their landscape-improvement money. I was mopping up an infinitesimally tiny fraction of a government budget to produce real and good effects, beautiful fields, full of highly various and valuable living things. Gaggles of schoolchildren would see this process at work and might, who knows, be turned by the experience towards the richness and solace that could be found in nature.

This Countryside Stewardship scheme, with its pompous and heavyweight title, may have been no more than a fig leaf. It represented only 2.5 per cent of a budget of which 97.5 per cent paid no attention to the meaning of the landscape at all. And even if the purpose of that little fig leaf, the scrap of green, was somehow to perpetuate the life of those landscape-destructive subsidies by distracting attention from them, did that matter? In the end I thought it didn't. We took the money, planted the hedges, managed the fields with floweriness in mind, had the children over, introduced them to beautiful fields, removed the eyesores and felt that this was public money being put towards an objective and public good.

At the same time, Sarah was beginning to address the garden. She is a woman who is not, it can safely be said, in any way moderate in what she does. With some relish she got the earthmovers in. She wanted the hill on which we were living resculpted. Most people might be happy to garden on the ground which God gave them but Sarah wasn't. Before we moved here, she used to spend a good proportion of her time in her London house rearranging furniture. In the course of one particularly distressing six-month period, the sitting room moved to three separate locations on two different floors, the various bedrooms and studies trailing woefully after it. I used to dread waking up in the morning. 'I wonder if the boys' room shouldn't be at the front?' the day would begin and I'd know we'd be in for many hours of the dining-room table being stuck at an angle on the stairs.

Now that we were on top of a hill in Sussex, the hire of large-scale earth-moving equipment was the outlet for this neurosis. It was becoming a spring ritual. As soon as the trees began to bud, I heard Sarah on the phone to Frisky Fieldwick, the earth-moving contractor. 'I would have thought two or three men for a couple of weeks would do it,' she said.

Frisky was one of the wildest-looking people I had ever met. He was the human equivalent of a Sussex wood. If you stood him in the undergrowth of some rather unkempt coppice, it would be impossible to tell he was there. Like those wonderful 1960s photographs of the model Verushka, her body painted to mimic the wall or trees against which she stood, Frisky would quite naturally have melted into his background. If I had to cast Puck, he would be the man. It might largely have been an effect of his hair, which looked like a hedge which has been driven through by an alcoholic with a brush-cutter, or maybe the jerseys the colour and consistency of a well-raked loam. I never saw him wearing anything except 10-gallon gumboots but Frisky was pure style. He would have looked fantastic on a cat-walk – Alan Bates meets Alexander McQueen – and he was blessed with the most spellbindingly seductive manner.

I tried to adopt something of the way in which he presented a bill. An enormous smile stole across his face, one eyebrow lifted a little, his hand moved like a gunfighter's towards his back pocket and you knew what was coming. 'You know what's coming, don't you, Adam?' he would say as the brown envelope, always folded in two, began to move elegantly towards you. The ritual continued. It was the same every time.

'I suppose you'd like it paid today, Frisky?'

'Well, we've all got bills to pay.'

'Yes, but have we got money to pay them? That's the question, isn't it?' I said.

'Some of us have and some of us haven't. I tell you, Adam, the money doesn't stay in my pocket. It'll be gone by ten o'clock this morning.'

I would write the cheque out to 'T.A. Fieldwick'. I had no idea what either T or A stood for. Meanwhile Frisky continued with the charm offensive. 'Now what about a pony for your little girls, a nice New Forest pony? They'd love a bit of riding round here. You've got just the place for it.' Frisky, apart from being earth-mover extraordinaire, also bred and dealt in horses, but thank God I managed to resist. 'Think of the memories they'll have,' he would say. I knew that he knew that I wouldn't dream of buying any horses, but we went through this little routine most weeks all the same.

'And why is it you're called Frisky, Frisky?' I then asked, going back to a favourite topic.

'Oh, it's because I was pretty frisky once!' he said, pocketing the cheque.

Paying for the earth-moving was the best part of it. For some reason, I found the actual digging deeply disturbing. Frisky would turn up in his Land Rover, which was loose and baggy, like a giant gumboot. With him would come his son Jason and Jason's younger brother, Ben. All three Fieldwicks had smiles like searchlights. They climbed on to the diggers and dumpers, Steve Moody, the garden contractor from up the lane, would join them, and the work began. Parts of the place, familiar not only to us but to the twenty generations of farmers who had thought of Perch Hill as their home and foundation, were sliced and eased away as if they had no permanence. This was surgery, landscape liposuction.

I couldn't bear to watch. I felt threatened and uneasy. I spent hours in my workroom reading yellowing copies of articles I had written twenty years before. Sarah was striding around like Patton in Normandy, buoyed up by change, by things happening, by the sight of the Fieldwick battalion making its all-too-definitive cuts.

I knew, in my rational self, that this work was a good idea. The area around the pond was a mess and needed improvement. I knew that and I was sure that what would emerge would be better than what was there before. But for all that, the process was troubling. The Fieldwicks were blithe, confident and skilful. And perhaps it was precisely that panache in execution which was troubling. I wanted to think that the place I saw around me was imbued with a permanence I didn't have myself. Everything we had done to Perch Hill Farm since coming here had been to enhance that sense of deepness and solidity. The house was becoming surrounded by gardens and garden walls. It was as though we were pegging the place down. These new earthworks were part of that process. When the little wood that we were planting on top of them had grown, full of hazels and hawthorns, wild cherries and one or two oaks, they too would embrace the buildings and farmyard, folding them in, diminishing the rather bruised openness we found on our arrival.

Sarah could clearly see the conclusion. I found myself stumbling over the way to get there. These machines were showing us how powerful we were in relation to the place. We could have demolished the whole lot in a couple of days. Within half a mile were the abandoned sites of two farmsteads which, until 15 years before, had everything you might have wished for: beautiful stone farmhouses, barns, yards, farm ponds, lives lived, memories treasured. They were demolished one day by the landlord, to prevent squatters occupying them. In the summer the nettles now grew there shoulder-high. It was that threat of erasure which alarmed me.

One day I was trying to show off to Steve Moody, the strapping gardener and dumper-truck driver. In Sarah's Blenheim-scale rearrangement of the local landscape, Steve had been acting as her *chef d'équipe*. Digging a couple of holes was meant to be my contribution. This was a mistake. Ever since an experience with a beautifully muscled masseur in the Andalucian quarter of Fez – I think the place was called the Hammam Ritzy-Sevilla – in the late 1970s, my back had not been what it should be. The big masseur, gleaming like an aubergine, as heartless as a stone, didn't seem to have cottoned on to the modern idea that 'No means no', He obviously thought my scream-groans were the English for 'Stop it, I love it'. Ever since, a niggle has lurked down there somewhere in my lumbar regions to remind me of the foolishness of youth.

But sometimes vanity and competitiveness get the better of me and I imagine that manual labour is something I can still do as well as the next man. That's why I walked out that day with spade and shovel to the site of the new orchard. Funded by a remortgage, a serious workforce had been on site for three weeks. A National Trust-standard car park had been installed. The ground had been cleared, what looked like blitz detritus had been spread over it, then a layer of stone precisely the colour of taramasalata, then some more blitz rubble of a finer grade, then blackish road scrapings, which might be mistaken for dead men's teeth, partly ground, and then the final Lutyensesque topping of rich, deep river stones.

Future archaeologists would scratch their heads over this centrepiece of the new works. How on earth could a hovel like Perch Hill Farm have deserved something modelled so carefully on the traffic-handling facilities at Sir Norman Foster's Stansted Airport? For the answer they would have to hire a medium to interrogate my wife's spirit. 'OBSESSION,' the Ouija board would spell out. I hoped the spirit world would understand.

My contribution to the Wealden Versailles was to plant and stake a couple of apple trees. The first was OK: hole, tree, stake, ties, earth, water, complete. The second went hole, tree AAAAAAARGGHHH, that neural, spasmodic, earth-shattering click, the moment every back-pain sufferer will recognize as the gates of hell. At least that was the sound inside. I wasn't going to show Steve anything was wrong. 'I think I'll go and get a glass of water,' I told him and broke for the house, stiffening, hobbling, crumbling upstairs and on to the bed, where I then stayed for a week, laid out, aged and with only my drugs and my laptop for company.

It was deep, deep agony at the time, an all-over clenching pain, but it wasn't long before I was under some really big-time medication. Sarah said I was as high as a kite but it didn't feel like that from the inside. It was the normal me in a rather good mood. Just a shot or two of morphine they gave me. Pure liquid wooze it was, straight in, happiness from a needle, and I felt fantastic. At last I'd come to understand the drug culture. The world was just beautiful, beautiful. Everyone loved me and I loved everyone. The sheep were woolly and the grass was green. I was having my own private Woodstock, but the sun was shining. We were deep into hippydom here and I was wearing my Afghan dressing gown. Modern technological medicine is a remarkable and adorable thing. Who needs the Tardis when you can have such a charming low-tech visit from the Burwash GP?

It was a dreadful/lovely time. Dreadful for the hint of paralysis it brought, those shuddering spasms of pain – the condition of washing as it travels through a mangle – but lovely as the drugs came on and you started to think this was the best of all possible worlds, that warm druggy dusk of sleepiness after pain.

I could see only the sky from my bed but I could hear life going on: trucks delivering yet further tonnages from the quarries of the globe; Sarah on the phone next to the open kitchen window saying, 'Of course I wanted 4,000. If I said 4,000, what do you think I meant? 400?'; Steve banging in post after post as his trees went effortlessly in; Ken Weekes's trowel clinking against the bricks of his new wall; the children screaming 'YES, YES' at the table football; the ducks on the pond; the ewes being penned up . . . Perhaps this was the kind of contentment bees felt, too, when the weather turns in the right direction. Who knows? Sarah came in. 'I've ordered the lawn,' she said.

'That's all right then,' I said. 'There'll be croquet in a matter of weeks,' and with those words fell fast asleep, dreaming of a heaven in which lilos covered the Perch Hill fields from hedge to hedge.

Transformations

I WAS upstairs, one spring morning, in my workroom in the oast-house, hiding. It was the first day of Sarah's courses. She was running them here at home. It was the first real attempt to make some money out of Perch Hill. She would teach people how to create a cutting garden, that is to say a garden not for looking at but for harvesting from. Seventy-two people, exclusively women, had signed up and, not of course that this was in any way important, paid up. But now Sarah had to deliver what her brochure had promised. I could hear her voice in the room below mine, telling the ladies about the need for 'a good friable soil and a warm, well-drained site'. An absorbed silence accompanied her words.

It had begun as a nightmare. The worst frost of the year had chosen the previous night to attack and we woke up to see the thousands of tulips and imperial fritillaries bowed and frozen, like ranks of collapsed ice-lollies. The euphorbia was drooping on to the path when it should have been all pert and Edwina Curriesque. Then Anna Cheney's car broke down in the village and so there was no one to look after the children. Half an hour later, the fleet of beautiful new silver Audis and dark blue Mercs started to nose cautiously into the farmyard. We knew there was nothing to pick. Secret panic reigned. 'Keep them inside till everything warms up,' I hissed at Sarah. A look of despair passed across her face. I then brushed the paths rather badly, hoping the ladies might think chaos charming. Then my father turned up and insisted on pressing his nose against the lecture-room window to see what the ladies looked like inside. God knows what kind of impression we were making.

By 11.30 a.m. we'd had the first session and then coffee. My sister Juliet cleaned the floor. The flowers had begun to lift in the sunshine. During the coffee break, two of the ducks decided to have sex in one of the flower-beds outside the kitchen. It was a terrifying vision of avian rape which finished only as the drake decided he'd had enough and walked away straight on up the back of his victim and then over her head, shoving it deep into the mushroom compost, as though her body was just another one of those things one has to negotiate in life. The rapee shuddered and shook herself like someone coming out of the shower in a shampoo ad. 'Do you always let the ducks into the garden?' a lady in an apricot cardigan said to me. Before I could answer, Rosie, in a deep Sussex accent said, 'Oh yeah. They eat the slugs, donay, Dad?' 'Oh really,' Mrs Apricot murmured and sipped her coffee, smiling a little distantly with her eyes.

Just before lunch, the party was out in the garden wielding scissors. Individual characters were emerging. Those who felt they shouldn't entirely destroy Sarah's incredible, sumptuous, intense and beautiful spring garden were hesitating before snipping one or two rather small tulips that were, if they were honest about it, slightly going over. Others, it was clear, thought they'd paid their money and so they were bloody well going to make their choice. One lady in particular returned with a bunch so large that you couldn't see her head behind it. I was hoping she would trip over. She didn't, but dumped her gatherings in a bucket and then said to me, 'Are you the man who tells everyone he writes articles in the *Telegraph*?'

'Yes,' I said, a little warily.

'Well, I've never seen any. Have you ever got one published?'

I could hear myself laughing and it sounded like the last drops of water draining from the bath.

Lunch was a rip-roaring success. Just before it began, Sarah had recommended that cut flowers needed one thing more than anything else: a good long drink. 'Oh yeaah,' one laconic American beauty said, 'and what about the clients?' Long drinks all round.

After lunch, the natural tendency was to wander a little among the flowers, over towards the edge of the garden from which the lovely view stretches down the Dudwell valley. That was all very well for background, but in the foreground, upwind in the prevailing breeze, was the totally failed, utterly disgusting and profoundly health-hazardous reed-bed sewage system whose accompanying pond still looked like a tureen full of mushy peas. It was of course inevitable that the entire course should end up surveying this part of the garden.

A Neapolitan smell of what are always called 'drains' blew across the ladies. I was praying they thought it something obscurely agricultural. 'Is that your wildlife pond?' one of them asked empathetically. A long and quivering moment of hesitation followed. Could I possibly get away with this lie? Which answer would be least likely to undermine Sarah's standing as a horticulturist of genius? 'Yes,' I said resolutely. 'We think it's very important to allow the wild a place in the garden and for all sorts of natural processes to be seen for what they are . . .' Grim nods all round at the wisdom of this, and I saw one of the Japanese ladies making a note in her notebook. I knew what it said: 'English character – never less trustworthy than when claiming high moral ground.'

By five p.m. it was over. We slumped around the kitchen table. A champagne cork lolled on the floor. Only another nine days like that one. It had been a triumph. Every one of them went away saying how much they'd enjoyed themselves. Sarah was exhausted but exultant. Everything worked in the end. I was thinking of the money and how to get the schmooze schmoozier. 'Ladies,' perhaps I would say next lunchtime, 'I'd like to show you our wildlife pond. It's so important to let the wild into the garden, don't you think?' Or perhaps not. Sarah said it might be better if I spent the day in London.

If I had only known it then, this was the beginning of Perch Hill's new and spectacular life, as a place in which Sarah's genius would come to flower. Alongside it, increasingly guided by Simon Bishop, the farm started to blossom too. He was the most inspiring of men, so attached to Sussex that he always said that if he left the county he would get a nose-bleed. But he wasn't stuffy about that, the very opposite in fact. As a lecturer at Plumpton Agricultural College, he was a forward-thinking educationalist. He was full of business ideas, he loved cows and cow-culture and did endless work to vitalize the Sussex farming people he lived among. Communitarian, blessed with managerial charms, a good friend, from the moment I met him I thought Simon was the model of the modern man, integrationist, optimistic and dynamic.

On holiday with his in-laws in Africa, he had been thinking about the future of the Weald, of places like this farm. How could we get things to work when the prices at the market were something of a joke? How to prevent the crude model of the capitalist economy slowly debilitating the Wealden landscape?

Simon's flash of inspiration was that farmers here didn't have to drown individually; they could float together. They could take advantage of their crowdedness. They could cooperate. Of course this idea is as old as the Weald itself. Every Wealden farmhouse of any age, made from the oaks that were cut from the surrounding woods, would have been made cooperatively. Here, as elsewhere in rural Europe, your neighbours would have built your house with you, and you theirs. Equipment and plough teams would have been shared. Favours would have been exchanged.

Common struggle would, at least in part, have dealt with common difficulties.

But that habit had faded. Modern individualism, in which every man on his own plot likes to make his own way in the world, had broken this system and perhaps by no coincidence redundancy was now facing this stretch of country. The ancient pattern of holdings which on average are about 90 or 100 acres was in danger of drifting into abandoned uselessness. The farms on their own were failing. Why not establish a local network of these farms? Many of them now, like this one, belonged to people whose main source of livelihood was not the land itself. As farms, they were no more than ticking over. They were in effect doing nothing and were ready – along with other large slices of rural England – to be steered in a new direction.

That direction, in Simon Bishop's mind, was what he called at first a Wealden Organic Network, WON. (It later became the Wealden Farmers' Network, because we didn't want to shut people out who weren't organic.) He would be overall director. I and the other landowners would enter into a profit-share agreement, based on the value of the resources we put at the Network's disposal. Arable land would count for more than rough grazing, a dairy herd more than a flock of sheep, a local high-street shop more than the kind of derelict shed I could offer. Because much of the land around here was doing virtually nothing anyway, beyond providing pretty views for its owners, the Network would not have a large up-front rental to pay. Profit-sharing would pay out only what had already been gathered in. The pooled acreage would be used for the rational planning of beef, dairy, cereal and vegetable production in a way that small and isolated farms couldn't manage economically.

It was a marvellous moment: so obvious, so clear, but providing the answer to a question I had been troubled by for years. How to make these marginal agricultural landscapes live again? The Network was to have its own shop and its own website, on which local availability of produce could be advertised, orders placed, days out on the farm announced, community picnics suggested, messages left and observations made. It could ask what the community might like to eat next year so that the right crops could be planted and the right animals raised. The shop itself would be in a village and would have a café attached so that its energy and example would spread rings of happiness around it. It would be a way of sewing together the very things which modern agriculture had severed: people and place, good food and good environments. When I heard all this I knew that Simon Bishop was the man.

So that summer, the Network was set up with three other local farms and the cattle arrived at Perch Hill. It was the moment I had been waiting for. Until then we had been a sheep place. The sheep had done what they were here to do: eat the grass, produce their lambs, mutely accepting their function as mowers with wombs, bleating from time to time, falling ill from time to time and dying for no known reason from time to time. The annual sheep cycle had established itself here in a neat and reliable rhythm but cattle represent something rather richer. It is a curious fact that the length of time a species has been domesticated by man is reflected in our relationship towards it. Dogs probably evolved into man's working colleagues about 13,000 years ago. Sheep followed just over 3,000 years later, cattle about 7,000 years ago and horses about 2,000 years after that. If you consider the relative wildness of each of those species now, the flightiness of their relations to us, the chronological sequence is clearly reflected in the animals' docility. Horses still have to be broken; cattle are still a little jumpy when you are in among them; sheep, at least the sheep you know, generally have a mildness and accepting calm which would be exceptional among cattle. But you would never consider having a sheep in the house as you would a dog. The shorter the time a species has lived with us, the wilder it continues to be.

To have cattle on the farm, then, represents a step further into the human relationship with the animal. It is something more equal than the keeping of sheep or a dog. For many years Simon Bishop had kept a herd of beautiful Sussex cattle at Ivyland Farm. They are, I think, the most beautiful cows in the world: a deep, rich, dark conker red, a cherry mahogany, often with a kind of burnish on the skin, particularly in summer when they lose the thick and tufty winter coat, replacing it with a pelt as sleek as a zebra's. It's a colour that goes outstandingly well with grass.

For years I had wanted to have a herd of Sussex cattle on this farm. They clearly belonged here. Until the invention of the tractor, much of the work on the heavy clay soils of the Weald was done with these cattle, pulling ploughs and harrows through the sodden winter fields, and the heavy carts on tracks that turned to fudge every winter. They are animals which have been bred over many centuries – there is a theory that the ancestors of Sussex cattle were among the indigenous breeds found in Britain at the Roman Conquest – for their docility with men, strength to cope with the exigencies of this landscape and what the Sussex Herd Book Society's history of the breed describes as 'the best fattening tendencies'. The Sussex is a miracle beast, 'second to none in early production of the finest quality of marbled beef – that excellent quality of flesh which will always command a better price than the cheaply produced or imported joints, which no butcher (mindful of his reputation) cares to place side by side with the rich juicy meat which the Sussex invariably furnishes'.

To begin with Simon brought 19 of his steers over to Perch Hill. They arrived in May, coming over in a truck from his farm at Netherfield one warm, nearly summer evening. The truck reversed up to the gateway. Its rear ramp was let down into the opening and 19 blinking, slightly nervy creatures came out into the sunlight. The steers had the last tufts of their winter coats still on them, a fog of their beautiful milky breath accompanying them out into the field. Their bodies then were light, slight little things, the sort of animals on which teenage bullfighters practise.

They were here just for the summer, happily grazing down among the buttercups of Great Flemings. I used to visit them: they surrounded me, licking my hands, blowing their sweet fleshy breath over me, nosing me from behind, curious and brave in a way lambs never are, recognizable immediately as individuals. Sheep always smell acrid, ammoniac, but these young Sussex smelled of grass and milk. I have in my hand the Standard of Excellence for Sussex cattle adopted in December 1907. For bulls, their skin must be 'mellow to the touch and covered with an abundant coat of rich soft red hair; a little white in front of the purse' – delicate word – 'is admissible but not desirable. A few grey hairs are not a disqualification.' They swish their tails and run their tongues up over their nostrils. You feel with them around you, much more than with a flock of sheep, the reality of the contract which herd-keeping man has made with these beasts. What can urban life provide that matches this strange cross-species intimacy? The 1907 expectation of the appearance of a Sussex cow is not only that her eyes should be 'bright and prominent' and the udder 'square, not fleshy' but that their general appearance should be 'smart and gay'. Who could resist that?

All through the summer, they were a wonderful presence on the farm. Mornings and evenings we would check them, either spread grazing across the hillside in Great Flemings, or clustered together in the shade of the chestnut trees at the bottom of that field, head to tail, the switch of one flicking the flies away from the nose of the other, a deep, lowering sleepiness emanating from the crowd. They were gentle animals. You could push between their big red flanks without any trouble. They would grunt aside, like sleepers in the same bed. Just a glimpse of them, away down in their own world at the end of the farm, would bring a smile to my lips. They looked as if they belonged here and the place looked better for their presence.

Slowly, as the summer went by, their bodies matured and deepened. They began to acquire that bullish silhouette, heavy in the front, where the brisket begins to bulk out over the breast bone, and across the neck where the tossing muscles thicken, but still quite light in the haunches. The mature animal looks curiously like a racing cyclist, hunched forward over the handle-bars, all his energy and focus on what is in front, and the rear legs, or rear wheel, nothing but a necessary balancing limb, useful in holding the front up, little more.

That's not how a butcher or grazier would look on it. You can measure an animal's readiness by the pudgy little bulbs of fat that appear on its hindquarters. The best cuts are there. By the beginning of November the Perch Hill steers were ready to go. The grass of the farm had fed them, nothing else. Simon Bishop came over in Sepember to have a look at them. He always arrived smiling, rubbing the palms of his hands together, a big challenging 'Hello' written all over his face, as if to ask, 'Right, so what new delight have we got to enjoy today?' We climbed over the gate together into Jim's Field and then walked down towards Great Flemings where the cattle liked to gather. The grass had turned green again with the first of the autumn rains and the cattle lifted their heads from it to look at us. A smiling satisfaction crept over his face. 'They've done well,' he said.

That autumn was a beautiful thing. The trees filling the valley looked like old velvet, crushed and rumpled, with different lights in different places. None of the garish polychrome of New

England here, or those sugary Japanese maples. An oaky, time-worn exhaustion, browning at the edges, coloured the view. I love this, the collapse from within, the life juices withdrawing as you watch. These woods are a forgotten library smelling of cigars. Silence hangs about them. There are leather patches on every elbow. The potpourri sits in blue glass bowls and the world is filled with the colours of sun-bleached tapestries. It is the last days of the ancient regime.

The strange, warm Neapolitan sunshine, in which the woods had been drenched for day after day, was interrupted by something else in the week the cattle left the farm. Sudden wind and rain smacked out of the west and thrashed around the hill. Our bedroom window was torn off its hinges and flung into the garden below. It looked like a car crash, some of the dahlias squeezed half flat by the one pane that remained whole, the others poking up past the splintered frame.

The entire arm of an oak tree fell into the lane, torn out of its stump. I found it lying half across a hedge. I cantilevered it up and shoved into the field. The surface of the lane was left littered with pieces of bark and brown leaves, like the floor of a workshop. Little twigs and torn pieces of tree were blown all over the paths. In the fields, the sheen of wet lay on the grass, like new silvering on a mirror. The tracks of night-time animals cut hesitant, wandering, investigative diagonals across it.

On one of those wild and windy days, the steers were driven up from the far end of the field, back towards the lorry from which they had come six months before. They weren't troubled or stressed. The Sussex is a biddable animal, and, one or two at a time, they clambered back up the ramp into the lorry. The wet day had left the hair on their backs sleek and rippled like a swimmer's. They smelled beefy now, where they had been milky before. They were going back to Netherfield and then within a few days to the slaughterhouse at Broad Oak. They were packed into the lorry and held in tightly so that the short journey would not trouble them. Ten tons of Sussex beef, easy animals, a good life, moved off up the lane. The lorry's engine struggled and groaned with the load. Perch Hill beef from Perch Hill grass. I could hear the weight of it, the best thing this farm had made since we came here. It felt like a signal of everything that might be good.

Through Simon, both Tessa, his wife, and a young Sussex boy, Colin Pilbeam, who had been one of his pupils at Ivyland, came to work at Perch Hill. Tessa helped Sarah in all sorts of ways, in her office, in the garden, running the courses and in the fledgling mail-order business we started to set up. Colin also worked in the garden, laying brick paths, making hazel and willow structures for the climbers to grow up, as well as looking after the animals, seeing every day that the sheep and cattle were all right, tending the dogs, pigs and the chickens, mowing the grass, weeding the beds, piling on the grit and the spent mushroom compost, sorting out the greenhouse, shooting the foxes, the rabbits and the occasional deer. Colin became, in other words, the invaluable core of Perch Hill. It is as much his as anybody's now. Every part of Simon's love for the animals found an echo in Colin. If a calf was poorly he would nurse it; the pigs learned the sound of his footsteps as he came over the gate to feed them; the dogs knew the note of his engine and would run out to meet him in the morning.

Even now, only a few years later, I am filled with nostalgia for those years at Perch Hill. Sarah was transforming the garden with her rampaging new appetite for intense colours, a kind of beautiful smoky richness that sharpened in places into the brazen and the garish. She was making something entirely unconnected with the Sussex landscape, but that could not have mattered less. I urged on her that the hedges at least should be of hawthorn, and if she was to have an avenue of trees, they should be hawthorns too, simply to root the garden in the place. But within the containers which those hedges made she should allow the exotic and the rich, the packed and the crammed in to have a dense, heightened life beside which the grass, the buttercups and the sorrel of my precious hay-fields could play their part as the rice to her curry, balm to her drama.

So she began to cultivate, above all, a sense of the large in contained places, a feeling for abundance as the source of delight in life, for colour as a visible symbol of a world beyond the ordinary. A vegetable and fruit garden was the inevitable outgrowth of a vision that saw the world as essentially consumable. It was all very well picking flowers, but what about lunch? And so, compartment by compartment, bed by bed, Sarah's grand demonstrative empire spread around the old farmhouse. The ladies came on her courses. The nature of the courses evolved, so that chefs and experts in meat began to come. A flood of adventure was sweeping through Perch Hill, with Simon Bishop, who loved Sarah, presiding over it all and Sarah's own appetite for the extraordinary providing the dynamo at its heart. Eventually, we thought that the garden itself might be opened to the public. It was only one day, a Tuesday in midsummer, but it had been written in our diaries for months, a glowing D-Day in whose service the entire year was shaped, a point in time beyond which all would be different.

The courses had brought a steady stream of people, but only eight or ten at a time and all very decorous. Once, I had made the fateful mistake of describing the arrival in our nettle-fringed yard of the blue Audis and silver Mercs as 'more welcome than the swallows in springtime, the life-enhancing sight of a fleet of cheques on wheels'. The phrase rippled around the neighbourhood like a computer virus, damaging Sarah's reputation and destroying mine. If I had been a computer, I would have crashed. Even nowadays unfamiliar women repeat the phrase with a look that is always the same: a roulade of offended morality, triple-layered with contempt, distaste and pity.

Despite these efforts at alienating the local population, Sarah had kept them coming through the doors. Each year, her business had doubled in size and with it the place had grown in variety and assurance. There was now a necklace of different gardens strung around the house and barns. Each was shut off from the wind with its own walls and hedges and so we were now living in a set of boxes where the flowers flowered and the fruit fruited.

By the oast-house the huge grasses and pompom alliums, the acanthus and tiger-lilies, the sugar-pink dahlias and rusty rudbeckias crowded the paths and overwhelmed the walls so that you had to push and creep your way between them. In the garden by the cow shed, blood-red cornflowers and marigolds filled the panels of a tapestry sampler. By the house in the kitchen garden, there is still a frost-resistant olive-tree and a shocking-pink bench next to lines of blue cabbages and lapping seas of nasturtiums. Roses flopped across the herb-garden beds and agapanthuses leant into the paths. The smell of a coppery fennel filled the air. Sarah's first cutting garden was, for that summer, lying mostly fallow, the ground cleared and plums fattening on the trees. The lower half, in its semi-neglect, was the most beautiful, a careless muddle of pink and orange poppies. When the wind blew they dropped their petals in a snowstorm of polychrome confetti. Across the track in the new cutting garden, a sudden summer thickness of annuals had appeared, all grown here from seed: tobacco and maize, sunflowers, larkspur and sweet peas, love-lies-bleeding and a beautiful lavender-coloured sage, which in the evening light glowed like blue lamps on green glass stems. Who wouldn't want to show this to the world?

The open day was in aid of St Michael's Hospice in Eastbourne, part of a programme they have of garden openings which runs for 12 weeks through the summer. Before the day itself, none of us here had any idea of the scale of the operation run by their fund-raiser, the dynamic Jenny Tyrrell. We were expecting 30, maybe 40 people to come. Even for them, though, there was some frantic tidying up: strimming at the docks, pulling out the bindweed, or at least anything of it you could see on the surface, mowing the bumpy bit of field we call the lawn, digging out the few remaining thistles, removing the lumps of sand that had hung around in various corners awaiting a purpose for the previous four years or so.

Perch Hill, incredibly, changed as I watched. Its old combination of mess and beauty, the ragged and the exotic, concrete and ipomeas, roses on the corrugated asbestos of the old bull pen, started to sort itself out, to shudder and shake itself, acquiring a curious and wonderful adult bloom in the process. I walked around thinking, 'Is this here? Is this, at last, the first hint of an arrival?'

The sun shone on the open day. Ladies with aprons arrived at nine o'clock. Rectangular cakes, each 3 feet long and 18 inches wide, were brought to the kitchen and left there under cloths. A tea urn was set up on the lawn. Men with sun umbrellas and fold-out tables with fold-out chairs arranged them as if at Ascot. A man asked me where the traffic signs should go. His lieutenants donned lemon-yellow Day-Glo jackets. We pushed the sheep and cattle into a distant field and shut the gate. At ten the cars began to arrive: 20, 40, 100, 150, more than 200 cars by lunchtime. Sarah put out her seeds for sale. Other ladies had brought pot plants, with which they made their own stall near the greenhouse.

The people wandered around the garden. A low murmur of enjoyment, like a human bee hum, came from within the various enclosures. They stopped and discussed the stranger plants in which Sarah indulges, the daturas and cleomes, pushing their noses in, flitting on, pollinating the place with their presence.

I couldn't keep the smile from my face. It was like watching an audience enjoying your own play in the theatre. All that preparation in private, that anxiety and stumbling, all that scrabbling and failing to catch the wave, and now here on this sunny day, the beautiful sight of shared pleasure, Perch Hill surfing on its own beauty. I met people who had known the place in the 1920s, the '40s, the '60s. 'You never told anyone the garden was like this,' they said.

Eventually it was over. Jenny Tyrrell counted things up: 618 people had come, £2,250 had been raised for the hospice, several acres of cake had been consumed. 'A triumph,' Sarah said last thing at night. She never says anything like that. 'Yes,' I said. 'Yours. It's yours.'

In order to supply the growing garden with the plants it needed, we put up first one and then a second polytunnel. It wasn't an entirely easy decision. I always used to think of polytunnels as plastic slugs littering the fields, the building equivalent of a rambler in a cagoule, but worse because immobile and stuck there for years. Beautiful buildings grow more beautiful the older they get. Polytunnels go in the opposite direction, declining rapidly with age, tarnished not burnished by experience, yellowing in the sun, flapping in the wind, turning baggy, loose-bellied and brown. A polytunnel seems to be dying as you look at it. They are made of the cheapest and most unseductive of materials: polythene stretched across a thin metal frame.

But Sarah was adamant. We couldn't afford a proper greenhouse (we had a shambolic lean-to of a greenhouse but it wasn't big enough) and the garden needed something more. I wasn't to sneer. The polytunnel would be the foundation of a new way of doing things. Rather than an ever-growing acreage of Tudor-style executive home estates with triple garages, she said, the polytunnel was precisely the sort of building of which there should be more in rural England. It embodied something that was valuable and important: affordable fertility. Nothing could be more valuable. So I listened to her, we found a site which was largely concealed by trees and we put one up.

Sarah and I and the dogs and the girls used to spend mornings in there, never more beautiful than in the early spring. Beyond the plastic membrane, Sussex would be pure hostility. A dilute sun seeped between the clouds and the wind cutting in like a slash of razor wire out of the east, pushing the cold damp of the Channel between the stitches of your clothes. But inside, within the polytunnel's embrace, it was a kind of early spring. The damp warm earth in the raised beds flavoured the air. The air itself was like a vegetable soup. Everywhere the little seedlings were poking up out of their compost nests. Nowhere was more full of promise on a spring day. The whole shape of the year to come was in here, all the vegetables and the tender flowers, one summer lunch after another, all in pure potential. 'Just add June,' the instructions on the packet should say. We spent all morning in there, watering the seedlings, tidying up, feeling the warmth which the plastic skin traps inside the tunnel. It is a way of enjoying summer twice, once in March when you see it ready, like this, waiting to happen, and once, you hope, again, when it's out there in the oast-garden and the vegetable garden, when each of these plants would be released into the flowering time of year, a flood of them running from the gates like children after school. It was the polytunnel which allowed Sarah to lay the foundations of her writing and gardening career. Here, year after year and season after season, she would sow the many different flowers and vegetables on a scale which would otherwise be impossible. It was our route to fertility. It was doing what this place and places like this are meant to do: growing, growing, growing. We should abandon the exterior view, consider places wonderful if wonderful things are happening in them, not if they conform to a surface vision of rural coherence. It was all part of our new confidence in life: the cattle, the beautiful fields, the Network, the people visiting the garden: all that made Perch Hill a happy place.

A Thick Pelt of Green

A DECADE after we had first arrived the farm was moving smoothly towards a calm and well-regulated condition. The sheep were better than they had ever been; we had got some pigs and some more chickens; the cattle were beautiful, at home and happy.

Slowly the herd built up until we had eight Sussex mothers, their daughters and their granddaughters. Calves were born outside in May or June on the fresh grass and then stayed with their mothers for the whole of the first year. Family groups were allowed to roam through several fields, choosing the corners they wanted to be in. Calves weaned themselves without stress. The family groups overwintered together in roomy barns over at Simon's farm at Ivyland. No additive of any kind was ever fed to them. It was beef farming as beef farming should be. When Simon discovered that Tesco required us to transport the steers live to an abattoir in Cornwall, he rejected the lucrative contract on offer for 'heritage beef' and had them slaughtered locally. Ten hours packed in the back of a truck on the motorway was not the sort of heritage he or I would ever have been interested in. And so they were killed down the road at the abattoir in Tottingworth Farm and then cut into joints, or the pig meat made into sausages at the Network's cutting room in Netherfield. People bought the meat direct from there, or occasionally at farmers' markets. It was Simon who brought a sense of system to the farm and to the Network, of doing the obvious thing well and calmly, of making everyone feel that they were lucky to be part of his life.

More than that, though, Simon knew, as his central vision, what farming could do to transform a young person's life. Many of the pupils he had at the out-station of Plumpton Agricultural College which he ran at Ivyland came from the poorest parts of Hastings. He made a particular point of looking after those with learning difficulties. 'What have you had for breakfast?' he would ask them every morning. Many would have had no breakfast at all and would eat nothing hot in the evening either. He set up a sort of field kitchen on his farm, in which many of them ate good, real food for the first time in their lives. These were teen-agers who had no experience of encouragement or sustained support either at school or at home. But Simon, often bringing them over to Perch Hill to help in the planting of a hedge or the shearing of the sheep, treated them with a generosity, gentle-ness and respect that I could only watch with awe. They loved him, clustering around him. And his method, slowly developed over the years, was to make sure that the pupils engaged with the whole production cycle, giving each group a small piece of land on which to grow their own vegetables and fruit, having them work on the farm with the pigs, collecting the eggs from the chickens, herding the cattle and sheep, gathering up the goods. It became the most wonderful destiny for the farm here: not a plaything, not allowed to go to rot, not being driven for the last penny it could deliver, but doing for these children what it had already done for me.

Under Simon Bishop's guidance, and with help from Angie Wilkins, who worked with him, we set about getting Perch Hill in even better order. First, it was some proper fencing, then dividing up some of the bigger fields so that the sheep and cattle could move on regularly to clean grazing and then piping water to each of those fields so that the animals could have a drink whenever they felt like it. It was a horribly expensive business: nearly 6 miles of sheep netting, more than 2,000 chestnut posts, thicker ones for the places where the fence turned a corner, diagonal strainers to strengthen those points, uncounted staples to fix the wire to the posts, gates, hinges, fastenings with which to close the gates. That was the hardware. Actually putting it up cost even more.

I knew we had to do this properly. I couldn't have looked Simon in the eye if we had scrimped or cheated. Anyway, I had begun to enter the farmer mentality enough to think that one shouldn't waste even a square yard of a field by putting the fence in from the boundary. So the boundary had to be cleared of all its brambles, overgrowth of hedgerow trees, old fence posts and old wire. Jimmy Gray, a fencer from Brightling, did the work, wading deep into the tangle with which our two biggest fields were surrounded. His tractor, with post rammer attached, got so bogged down in the glutinous fudge that lurks an eighth of an inch below the grass, that he had to summon an enormous tracked digger to get him out. The digger could get no grip on earth which had the consistency of Angel Delight. At one point it began to slide unstoppably towards the lip of a wooded bank, the driver incapable of doing anything to save himself. Only by slamming down the machine's telescopic arm and jamming the armoured tip of the yard-wide bucket into the earth did he bring himself to a halt. The field itself looked half-ploughed, but we disced and rolled the following spring and you would never have known the difference.

A farm doesn't work without rigour, without doing as much as you can. It's all very well for developers developing housing estates, or borough engineers devising road schemes, or great men in their great houses laying out their parks, to think that a little spare ground can be lost or used for nothing, for effect. But not a farm. The treasured qualities of the English landscape – or any closely farmed place from here to Java, anywhere there has never been much fat to spare – come from this squeezing of the resource, this tightness in the use of land, the precise interlocking of field and wood and track arranged as closely as the words in a sentence.

So, slowly, that is what Jimmy did and we got our new fence, about £25,000-worth of sheep-netting, half of it paid for with a grant, with 27 new gates at various points through it, the whole lot enclosing about 70 acres of the grazing on the farm.

What Jimmy had done looked efficient but there was a certain rawness to it, where the hedgerow trees had been lopped and all the old field margin scurf cut away and burnt. I knew all that would soon return. The brightness of the fence posts would dim and the fields would once again acquire the aged, patinated calm which all this work seemed to have erased. We hadn't sprayed anything and all the woody hedge plants were still there. It was not as if any permanent damage had been done. Behind the fences, we were planting new hedges and filling up the gaps where the hedges had decayed into isolated fragments. In 10 years it would be all right.

In fact, this work on the ground, making the place feel right in itself, redeeming something which had been on the way down for too long, turning that trajectory around: all of that was a deeper pleasure than simply enjoying or longing for the aesthetics of the old. The following summer, in the heat of the morning, with my shirt off and the sweat running down me, I did a job that had been hanging over me for months. The wood which we had cut from the field margins had been lying there in lumps all spring. At last, I was getting it in.

'There is little better in the world than this,' I thought. I had my small John Deere and its trailer next to me. I was loading it with the lengths of wood from the field. The Sussex steers were looking at me like a panel of jurors confronted with a thief. The county after which they are named was going blue in the distance as it dropped away from our own green fields. The swallows were skimming the grasses like swimmers after the dive, seen beneath the water, their hair slicked back on to the bones of their heads.

I was taking in the lengths of wood to burn on our winter fires, gathering our winter warmth in the middle of the summer heat. I am not quite sure why this felt so good. Perhaps it was simply the pleasure of planning, of doing something now which meant that we would be prepared when the winter came. I should have done it months before, before the grass and thistles grew up among the piles of cut wood, but other things had intervened and I hadn't. Now, at least, the ground was dry. The place by the gate where the water had gathered in little ponds and which was filled for a week or so with frogspawn was now a dusty hollow. The wheels of the trailer bounced happily across the dry surface and the steers' hooves drummed on the field as if on a wooden chest.

This was more, though, than complacent pleasure in my own timeliness. I love, above all things here, the connectedness and lack of profligacy in these things: new fence pushed to old boundary meant new firewood for old fires. It was maintenance, that word which, as somebody reminded me, means, etymologically, 'holding in your hand', the Latin, I suppose, for 'keeping up'. In doing this, in loading this trailer, in sweating over the wood, and then in splitting it outside the wood shed, with the heavy head of the splitting axe, cleaving between the fibres of the tree, I was holding Perch Hill in my hand, keeping it up, repeating the pattern.

It was not the life for lying back. Sarah's courses were doing well. She was writing more, increasingly both on gardening and food, and the buildings we had weren't good enough for what had in effect become a gardening and cookery school. It was an exciting and terrifying time, as if we had left behind our childhood and entered a fiercer and more adult world. Anyone who has set up an enterprise like Sarah's knows, above all, the difficulty of making it grow. I think we fell for every mistake in the book. We tried to run a mail-order business from the barn. We installed more telephone and computer equipment than I ever want to see again. Her customer base ballooned from 300 to 25,000. Delivery lorries crammed themselves down the lane. We employed far too many people: at one point there were 35 of them coming to work at the farm. We borrowed too much money. The tiny beginnings of the business, in which we had hand-coloured her seed catalogues on the kitchen table, got left behind in a welter of largeness, other people and alien ideas. It couldn't stay small: nothing can. But as it got bigger, it seemed to acquire a vast, dominating presence in our lives, like a puppy transmuting into an elephant. People we scarcely knew were making decisions about what 'Sarah Raven' – 'Not the person, Adam, the *brand*' – really stood for.

On open days, quite regularly, 1,000 people would arrive. The cars parked in the Cottage Field stretched for hundreds of yards down towards the wood. I sat at the gate taking the entrance money. Colin directed the traffic. Neighbours came in for free. Sarah gave talks in the old kitchen, Tessa Bishop ran a small shop for seeds. Bea Burke, the young Hungarian whom Sarah had taken on as a gardener, sold the plants she had raised in the polytunnels. Simon Bishop and Angie Wilkins sold cuts of meat and sausages from the Wealden Farmers' Network and other local entrepreneurs – nurserymen, bakers and wine-makers – sold their wares in a little market outside the barn.

The old 1940s cow shed, in which I had lambed the lambs, had become little better than a dusty junkyard for most of the year. We decided to convert it into Sarah's new school. I had met the Scottish architect Kathryn Findlay and together with her team we designed a beautiful new building: a greenhouse clamped to the southern side of the big old shed, the asbestos roof replaced, offices and a kitchen all squeezed in. Four separate doors connected the greenhouse to the school so that life could flow happily from one to the other. Everything inside was to be white and clear so that the flowers themselves – and the delicious cooking smells – would become the stars of the show. The big bank outside the greenhouse was planted up with vegetables and annuals and from that greenhouse the fields and woods of the southern part of the farm seemed to spread out like a ruched and ruffled apron of interleaved greens. 'Look,' the view said. 'Look how wonderful Perch Hill is.'

The new school became the most beautiful building at Perch Hill. Nipper Keeley made a giant maplewood table to go in its greenhouse, and when the courses were not going on, we took to living in there, half-amazed at the little slice of glamour which the building had brought into our lives. Matthew Rice told us all how to keep chickens. We became very nearly self-sufficient in meat, eggs and veg and, away from the chronically troublesome business, Sarah found all the solace she could hope for in the burgeoning productivity of her garden.

So this was the situation in 2004: we had done well with the farm at Perch Hill. It was in good shape and in good hands. Sarah had made something astonishingly beautiful in the garden here, completely of its own kind, a garden dedicated to the idea that its beauty consisted in its own abundance, a place in which nature could be allowed and encouraged to be its most demonstrative. Her garden was a theatre of fullness. On summer mornings I would walk out into it – heady, thick, dripping, colour-drenched – and feel that she had made a miracle, a descendant of Great Dixter in part but heroically itself. We had done well by some of the buildings but others were still in a sorry state. We had made a place our children loved.

The farm had settled into clarity and simplicity. On a summer's day I took one of my farming neighbours to see what I think of as our most beautiful field. It's always been known as the Way Field, but we call it Rosie's Field, because it is where, during our first summer here, we had a party for her first birthday. We put a tent up next to the top hedge, its flaps open to the field. Sarah had decorated the eaves of the tent with the looping lines of a summer swag: poppies, cornflowers and marigolds all picked from her new annuals garden, providing a frieze of colour above Perch Hill's most seductive view. The 11 acres of standing hay spread out in front of us as an apron of red sorrel, with yellow and blue vetches, the sharp, almost blue-white heads of the yarrow flowers, the grasses going tan and khaki, the lilac tufts here and there of the flowering thistles and the pools of red and white clover. Beyond them – and this is one of the best things the summer does to the Weald – the sun was casting deep eye-sockets of shadow along the edges of the wood, in which only the trunks of the birches stood out as stripes of paleness.

The party was the first time we showed our friends what this place was like. It was more a demonstration to them than a party for Rosie.

I felt, rather absurdly, proud of what you could see from the tent; of what I could now say I owned, with its butterflies dancing in and out of the hay tips, and the way it so clearly fitted the picture of what a place like this might be. When the wind turned in the right direction, the smell of the field blew over us, the smell of warmed vegetation. It also brought the sound of the bells of Burwash church. Down in the bottom of the field, a troupe of deer processed in a line across the field like ducklings at a fairground shooting arcade. Afterwards, we lay down in the hay, careless of the extravagance of this, and looked up at the sky through the rim of stalks.

Now, five years later, I was lying in the field again, writing this. If anything, the hay was better, more flowery than it had ever been. More vetches made blue patches in the bleached grass tops. The yellow rattle I had sown over the previous two winters had now taken in dense colonies 10 or 15 feet across and was setting its own seed. I manage the field for floweriness: a very late cut of the hay, not before the end of July and sometimes later; grazing the aftermath so that by midwinter the sward is right down to nothing and in places small patches of bare earth are visible between the plants; taking the sheep off so that they don't hammer it too hard; then allowing the cycle to begin again. As a system, it seems to provide just the conditions on which the flowering plants thrive. The dogs were now lying beside me, panting in the heat, as the late hay threw its diagonal, muddled shadows across the page.

Despite the Elysian nature of this, all I can think of, and I know why, is what my farming neighbour said when I showed him this field last haytime. Its polychromatic shimmer moved under the wind coming down the valley. The dogs bounced around in it as though in a playground, disappearing from view momentarily at each bound. 'What do you think of *that* then?' I asked him. This wasn't even a negotiation. I wasn't trying to sell him or persuade him of anything. I just wanted to have him agree that the Way Field was a marvellous thing, in the way you might show someone a picture you liked or tell them a story that had intrigued you.

'*What* about that, then?' he asked, killing the enthusiasm in the question, shifting the emphasis from 'that' to 'what'.

'Isn't it a marvellous sight?' I repeated, pushing for a yes.

'It's a desert,' he said. 'It'd hardly be worth baling.' We walked out into the field, pushing past the knapweed, sending up the meadow browns, like the sparkle above a glass of lemonade. 'It's all top and no bottom,' he said. That year, when the spring and early summer had been dry, he was right. What looked like a solid slice of grass, 15 inches thick and 11 acres in extent, turned out as you walked through it to be made mostly of air. The tractor-borne mower would be spending most of its time slicing on nothing.

Ever since, I have looked for bottom in a hayfield. You can't judge it from a distance. Only if the understorey of grass clogs your shoes as you push through it will it be any good. This year, the deep reserves of damp from the wet winter and spring had ensured that the Way Field, beneath its tawny, flamboyant upper layer, was thick with goodness, a thick pelt of green which soon enough would make many hundreds of bales. In a way the farmer was right. But if I understood he was right, I didn't and still don't accept that I was wrong. For the Way Field to be beautiful is a form of value too.

But all was not perfect. Increasingly, like an infection coming in from beyond the borders of the farm, we had been having a problem with ragwort. You have to pull it, one by one, and the result is a peculiar mixture of pleasure and pain. The pleasure is in doing it at all. Walking through the hay meadows, just a moment or two before they are to be cut, with the long distance of the Weald running away to the blue hills above Rye and what I always imagine is the brightness on the horizon there from the sunshine reflected off the Channel, the dirty, egg-yolk yellow of the ragwort flower-heads stands out here and there among the tan of the grasses.

Ragwort is the nastiest of all English weeds, a British native which our imperial past has exported to the rest of the world. When we sent our cattle to the colonies in the nineteenth century, feeding the beasts with hay in the holds of the sailing ships, there must have been ragwort seeds in amongst it. The cattle ate the seeds, walked out on to the pristine pastures of the new world and deposited them there in a beautifully fertilized pat, inviting the ragwort to flower on a new continent. If you look up ragwort on the Internet now, a low moan of despair comes back at you from those former colonies. In New Zealand and Western Australia, in Nova Scotia, Oregon and Washington State, ragwort is booming, threatening to overtake thousands of acres of pasture and with the prospect of doing real economic damage.

It is a killer. Even as you touch its tall and horribly vigorous stems, a kind of chemical, pollutant taste comes into your mouth, catching at your throat. No one could ever describe that smell as a scent; it is more of a sense-attack, the olfactory equivalent of a thistle, a spike-cocktail of sourness and unattraction. Don't touch me, the airborne message communicates. Leave me alone, let me flower, let me thrive. I am not on your side.

Usually, unless they are particularly hard-pressed – which can happen in the spare, dry grazing of the western United States but only rarely in England – stock will not eat ragwort that is still green and growing. The real danger is if the ragwort is cut. Either dry in hay, or if mown off, or even bundled wet into silage, it becomes more palatable to the animals. But if it is a little tastier, it is no less poisonous and an animal that has been feeding on ragwort is in mortal danger. A cow or a horse which has been exposed to these toxins is a disturbing sight. Such animals lose their appetite and start to chew on fences. As it worsens, they begin to stagger, go blind, bump into obstacles and eat dirt. Hair falls out of the manes of horses and their hooves start to flake away. These are all the symptoms of liver damage which within two or three weeks, if they consume enough of the poison, can kill them. The damage to their livers is irreversible. The only thing a vet can do is shoot them.

We didn't have that bad a problem with the infestation ourselves. Until about 2000, I don't remember seeing a single ragwort on the place. Then there were one or two, which I pulled up and burned, and now the plants are appearing just here and there, little yellow spots of pointilliste horror scattered around the farm. I don't like it. I walk around the fields pulling them up, tolerating the chemical taste that soon starts to feel raw and thick in the back of my throat because I know that to do this is at least ensuring that this year's hay will be all right. By pulling them and then burning the bundle I have gathered, I can at least be sure that these plants won't set seed. The horribly vibrant flower-cluster on each of them, a mountebank coarseness to it, like a carpet-bagger rolling into town and setting out his wares, can, I am told, produce 150,000 seeds, of which two-thirds are likely to germinate the following year. Unattended, it would not be long before the farm was awash with the things. In Oregon, damage done by ragwort to livestock in the 1980s was running at about 4 million dollars a year and whole counties were becoming unusable by graziers, a low flood of poison creeping across the productive landscape.

When you pull it, the plant comes smoothly and satisfyingly out of the ground, a big lump of root attached to its foot. It reminds me of that wonderful moment when you were a boy and at last a wobbly tooth would come oozing out of its socket and you were left with a strange fresh soft gap in your gum into which your tongue would dive as if into a pool.

But even as I do this, I know I am not doing the right thing. My pulling of this year's plants will, inevitably, leave small pieces of root in the ground and ragwort will resprout from any piece of root, however small. This is the quick and lazy way to do it. If I were not committed to an organic system which uses no chemical sprays, I could attack them with a herbicide called 2-4D. But I am, so I can't. What can I do? One is told in this country that the only possible thing a landowner should do – and any of us with any ragwort whatsoever on our land can in theory be served clearance notices by the Ministry or the local authority – is to dig them out. Even those who have sprayed off the tops are told to dig out the roots.

That might, I think, be possible on our land, a slow and patient working through the fields, attacking the ragworts one by one: immensely slow, immensely old-fashioned, immensely expensive in terms of the labour required. But you have only to take a drive through England in the summer nowadays to see that the ragwort is out of control. Thousands of acres of scarcely used ground and motorway verge are thick with it. Ragwort is every-where. It is rampant and almost totally unchecked. Is there any point in my trudging round with the dog at my heel, pulling out the weed on our few acres when the rest of the country was going to a form of toxic hell?

Of course, that has always been the big question hanging over Perch Hill. If the big systems are so wrong, is there any justification for making and protecting a little slice of delicious paradise here?

Sometimes I long for the techno-fix. If only, in one seamless sweep over the ninety acres of this farm, we could swish through with our weedkiller spray, ridding ourselves in one gesture of everything that hurts and troubles us. It would be heaven in a boom-sprayer.

The sow thistles, creeping thistles, docks, ragwort, nettles, brambles, dandelions and creeping buttercups would all meekly accept their fate, standing there mutely as the chemical rain fell about them, drinking it in, absorbing into their roots the substance that would do for them. I would never need worry about the weed problem again. Without anxiety I could take guests for a walk across the fields, without the detours through patches of woodland in order to avoid the worst and weediest patches of field.

But of course we couldn't zap them because we were organic. And why were we organic? Because we were. As I wouldn't have accepted that reply from my three-and six-year-old daughters when I asked them why they were painting the dog's ears, I could hardly offer it up myself. Why don't you tidy up with the best tool that modern technology can provide? people ask me. It is as if you were using a broom to sweep up the sitting room and refusing to use a hoover because to do so would consume valuable fossil fuels and add to global warming. It is eccentric to the point of lunacy.

Is it? Glyphosate, the world's biggest-selling herbicide, is seductive. It leaves no harmful residue because its active chemical either clings to soil particles and is neutralized or degrades in the air. And because it hangs on to the soil, it doesn't leach out into the watercourses. It does not harm people or animals because the mechanism by which it kills plants acts on a biological system (called the shikimic acid pathway, whatever that might be) which exists in plants but not in animals.

So, the chemical crew says, what is wrong with glyphosate? It's the perfect killer. Why not go for the full zap? Use it with a machine called a weed-wiper. This ingenious mechanism is a roller on wheels which is towed behind a tractor. The weedkiller is spread along the roller by a series of little nozzles. The clever part is that the chemical-drenched roller rolls not along the ground but a foot or so above it. You put your sheep and cattle in the field. They graze away at the grasses (and wild flowers) but they leave the noxious weeds. After a few weeks, those weeds are left standing a good foot or so above the sward. Out with the animals, in with the weed-wiper, set at a height which will spread the herbicide only on those tall plants it touches. Result: end of weeds, protection of everything else. What have you got to say to that?

There are some techno-answers to the techno-zap. Glyphosate reduces the amount of nitrogen which clover or beans can fix in the soil. This beautiful mechanism, by which a plant draws the nitrogenous goodness out of the air and stores it in little nodules on its roots, building up the fertility in the grass years which is then made use of in the arable years, is the hidden mechanism on which traditional and organic farming rely. If you start using glyphosate, you start eroding that all-important cycle. The other little things that inhabit the soil don't like it much either: bacteria, fungi, yeasts and some invertebrates suffer in a glyphosated environment. The parasol mushrooms I had been picking up from Great Flemings in the autumn, and the extraordinary puffballs which proliferated everywhere on the farm since I kicked an old dead brown one around the fields one year, puffing out its brown smoke like the dust on chocolate truffles – none of that would thrive if I was happily weed-wiping away.

I never wanted to rely on the techno-answer. Instead, we top our weeds, year after year, three or four times a year, with a tractor-mounted topper, just cutting them down whenever they perk up. Fred Groombridge used to do it, and then Peter Pilbeam, whose brother Colin had come to work in the garden. It is most important to do it in the autumn, sorting the place out so that it doesn't look stubbled for the long winter months to come. That involves allowing the weeds to grow so that we can then chop them back. You have to contemplate what is wrong for a few weeks before doing anything about it. But that is precisely what is good about doing it this way. Topping is not a final solution. It doesn't zap for ever something which you find unattractive. It is part of a continuing process, not a sterilizing or eradicating but a managing and an accommodating.

After the chaotic beginning, described in the earlier chapters, we started topping effectively in 1996. It never seemed to do that much. I remember one year walking across the fields with Christopher Lloyd, who had come over from Great Dixter on a summer evening. He and Sarah had talked about the new way of gardening, large and bold, no reticence, no sanctified sweetness, everything demonstrative and large, the deep bloody soaked-in colour of the dahlias, banana plants where Lutyens had made a rose garden, his thinking of a garden not as a place of reticence and delicacy but designed for flamboyance, grandeur and imposition, as big a presence in your life as a power station. Watercolours were so passé; this was opera.

I loved him for all of that, his twinkly, often rather cruel naughtiness, his knowledge that we were creaking along on scarcely enough money, his unstoppable truth-telling. So I loved it when he said how beautiful the fields were at Perch Hill, their long blue slide to the distance, their Sussexness. I loved it as much as when someone said how lovely my children were. And I knew that Christo would not lie about such a thing.

All the fields here, hedged or ploughed, grazed or not, are all, to a stranger, or to anyone who is indifferent, pretty much of a piece and pretty much the same as other fields on other farms where sheep and cattle graze, where the oaks grow, hedges are cut each sodden March and the hay made each burning July. They are part of what strangers call a 'landscape', that distant word no one ever applies to a place they know but which drapes the convenience of singularity over every local reality. Landscape; prospect; scenery; view: synonyms for ignorance.

Those words dissolve when you know a place. Landscape-with-knowledge becomes its constituent parts, breaks up into farms and woods, tracks and rough, damp places where rushes grow and where, after rain, water collects in little grass-drowning pools. That is the level at which I know this place. I know what every notched corner looks like at any time of year. I have in my mind a houseful of seasons for every cranny: the future presence of a splash of garlic by the fallen oak on the sunny edge of the bluebell wood; the future presence of those bluebells collapsed and flowerless, as though a passing flood had flattened and drowned them; the future rattle of the acorns sprayed by autumn winds and beating through the trunks of the chestnuts or hazels beneath them; the future absence of any single thing to lift the spirit in the deep and rain-soaked blackness of the winter trees. The whole year co-exists in a known place.

These layered changes in the seasons, this mixture of frag-mentary preservation and partial erasure, comes near the heart of a place's significance. Each place, each evolving corner, is a form of current memorial, a marriage of life and permanence, halfway between now and then, between the made and the given, the local and the abstract, beautiful, in Ivor Gurney's word, for their 'usualty'. That is why I didn't want to poison it.

So when Christo said, 'The only trouble is the thistles, isn't it, Adam?', looking up into my face, like a small boy saying something rude and true like 'Your breath smells', it was like a dagger in my heart. For all the *look* of Perch Hill, its landscapy virtues, the underlying truth was in the thistles, that terrible crunchiness underfoot, the inability on a summer evening to lie down in the grass because the grass wasn't comfy. I explained to him what we were doing, the repeated topping, the year-in, year-out nailing of the thistles, their year-in, year-out return like the most reliable of crops, but the sceptical look didn't leave his face. 'Chemicals, Adam?' he said.

Then, over two or three summers, from about 2003 to about 2006, and in a process that is still not quite complete, something extraordinary happened: the thistles disappeared. In field after field, the grass became a picture of what grass is meant to be. All through Jim's Field and Great Flemings, in Beech Meadow and Rosie's Field, a pure, smooth sward appeared, made largely of grass, a lot of buttercups and large amounts of sorrel and red and white clover. Why? Was it because we had been bashing away at the thistles with the topper for years and they had finally succumbed? Or the way we had grazed it some winters very tight because we didn't have quite enough grass for the stock? Or was it, as some people suggested, a new thistle disease which had come into the country from Canada and was slowly eradicating them, first turning them pale where they grew among the grasses and then killing them off entirely? I had certainly seen quite a few strangely pale thistles like blanched asparagus standing in the grass.

Whatever the reason – and it was probably a mix of all three – it felt and still feels like a miracle. Our hay is now prickle-free and the cattle in the winter can plunge their noses into it without any of that hesitation with which they approach prickly hay. The surface of the fields, when grazed, is a smooth, shorn carpet like short, well-washed hair. And when we shut up one or two fields in the summer to grow hay for the animals' winter food, the flowery grass is thistle-free, an unbroken expanse, eleven acres in the Way Field, fourteen in Great Flemings, of what we wanted to be there. I cannot tell you how deep the pleasure of that now is to me, walking through the long grass, hoping against hope that I wouldn't find the thistles there, lurking in the camouflage of buttercup and sorrel. They never are. There is no reason, Angie Wilkins says, they should ever return again.

What is more, we have done this without chemical poisons. We have done it by looking after it, by managing it as it needed to be managed. I never thought it would work, but it has. And that, mysteriously, turns out to be one of the consolations of time passing here. I think somewhere deep in my preconceptions was the idea that things will decline, that as the world turns it slows down, that what will come can never be more than a sad and diminished version of what came before. That all you can do is hope to recover some of the glory of things that are largely gone, that at Perch Hill there might be a fragment of the world which is still good and whole. That climbing part of the way back up the slope is the best you can hope for.

But here was something that flew in the face of that nostalgic gloom. In the deepest possible way, in the deep networks of the fields' ecosystems here, we had made things better, not in some flash, applied solution but in the reality of this place's nature. I still cannot quite believe it. We haven't brought anything into these fields, nor built anything. But we have somehow allowed them to become themselves. We have allowed them to shrug off the thistle burden which probably came here in an attempt to make them more productive in the 1960s or '70s. The thistle seed might have been mixed in with the seed of the new rye-grasses that were drilled here then. We don't know what it was but now it has gone and the fields of summer grass look well nowadays, with a smile on their face, as if the farm has been made good again.

Feeding the Sensuous Memory

In early 2004, my father fell ill, suffering from what turned out to be his last illness. We went to live with him, at Sissinghurst, in the house I had grown up in thirty years before. There had always been a strange parallelism in the relationship of Perch Hill to Sissinghurst. They were not that far from each other, a little more than 15 miles, one just in Sussex, the other just in Kent. The geography is pretty much the same in both places. Perch Hill's bit of Sussex is a little rougher and woollier, the hills are steeper and the valleys sharper.

In essentials, or in the kind of essentials a boy would know about, the places are the same. The clay is just as sticky, the streams run similarly in deeply cut and deeply shaded runnels through the woods and out into the fields, where on burning days they become beautiful cool trenches between the parched crops on either side. Kingfishers fly the length of them in both counties. The pattern of fields is the same, the bocage which makes each corner into an individualized world. And above all the woods are the same, still coppiced in both places, still with their beautiful oak standards remaining from cycle to cycle as the underwood is cut from around them. All this was what I loved more than anything else about home when I was a boy. Much of it is what I had responded to when coming to Perch Hill.

It might seem that everything we were doing at Perch Hill, getting the farm up and running, restoring the buildings, making a garden, attempting to turn the place back into what it should be, not just a place to sleep and eat but a centre of life, of people coming and going – that all this was driven in me by a need to remake the place where I had lived when I was a child.

But there was something else in play here. At Perch Hill we were trying to make a version of home which was truer and better than what I had known as a boy. This wasn't simply a repeat; it was an improvement. And the improvement was not one of status or importance. The garden at Sissinghurst had been created by my grandparents, Vita Sackville-West and Harold Nicolson, in the ruins of a great Elizabethan house. It was filled with echoes of grandeur, faint and crumbled maybe, but which Perch Hill could never hope to match. No, the improvement was at the other end of the scale. What Sarah and I found at Perch Hill and everything we have done there has been driven not by any sense of status or importance but by a need for rootedness. We both wanted to make a place that felt richly itself, that glowed with its own life, that felt as if all its ingredients were contributing to a small but particularly whole economy, in the Greek sense of that word: *oiko-nomos*, the law of the house. It was to be a place that felt good in itself and good to be in. Perhaps the drive to make a place like that came from a certain lack at Sissinghurst, a hint of well-delivered sterility, a smartness, a sleek swishness which concealed an emotional thinness, the absence of a beating heart. That in the end is the subject of this book: the search for a beating heart.

So Sissinghurst and Perch Hill came to play different and complementary roles in our lives. My father died in September 2004 and we stayed on at Sissinghurst then, yo-yoing between there and Perch Hill, sometimes sleeping in one, sometimes the other, working in one or the other. Sissinghurst was the most heart-stoppingly beautiful of places but it came with tensions and burdens: a National Trust place; many people who thought of it as theirs and us as interlopers; and a lack of control over the garden and the environment which at times felt as though we were trying to live in a glass-walled museum. On a couple of occasions we decided we could no longer live at Sissinghurst. We had tried for long enough to remain buoyant in the face of a certain bleakness and unfriendliness from a couple of individuals. We had done our best to give Sissinghurst what we could. But it is impossible to live somewhere in which you feel, every day, from some quarters, resented and unwelcome. It made Sissinghurst toxic.

A strange moment came in the summer of 2006. The house at Sissinghurst needed to be rewired and so we moved back to Perch Hill. It felt as if we were repossessing a simpler past. Perch Hill was unpolitical. There was no need for manoeuvring there. It felt like home. We yawned and stretched with relief, the muscles of our relaxing selves suddenly finding room where they had been cramped into a box before. I don't think we had realized how tense we had been. That evening was warm and we went out with a picnic on to the farm, to the Way Field, where the hay had just been made and turned and was lying in rumpled rows, waiting for the baler the next day. We played there with the girls. I made a hay castle for them, and a hay hurdle course, like the Grand National, and then we all sneezed and drank beer with our bread and cheese. I felt as happy as a happy dog, as if I had been let out and let off. I remembered something Gaston Bachelard, the great French philosopher of place, had said: 'The purpose of a house is to allow you to dream in peace.' I lay out in the warm starry night that evening, after the others had gone to bed, and rolled that phrase over in my mind. It wasn't a house exactly which provided the shelter here. It was more a form of psychic skin, a cloak, a tent, under which you could hide, and somehow in that hidden condition feel the world as it was meant to be. How at Sissinghurst, a busy, multi-faceted public institution, could one ever hope for that? But here at Perch, under this roof of stars, nothing could be plainer. Everything Sarah and I had once seen in this field was still here.

But we did not stay at Perch Hill. When the rewiring was completed at Sissinghurst, we went back there. The truth is that Sissinghurst had diverted us. As I have described in another book, I took it upon myself to persuade the National Trust to re-establish some of those qualities of authenticity and rooted vitality which had withered in the place since I had known it as a boy. A half-acknowledged conversation was going on in my mind. The Sissinghurst I remembered from the 1960s, with its hops, orchards, cattle, pigs and chickens, had been intuitively or not the basis for what I wanted at Perch Hill. Those huge communal lunches we always had in the Perch Hill kitchen in the early days, when Will and Peter Clark, Ken Weekes and Sarah and I and Rosie and sometimes Steve Moody (the garden contractor) and his wife Alison, and Anna Cheney (the girls' nanny) and her daughter Charlotte, when we all sat down together in the kitchen to a mountain of pasta or two or three chickens – what were they but a kind of recreation of the thick communal well-being I had sensed in the air as a boy at Sissinghurst?

But if Sissinghurst had shaped Perch Hill, Perch Hill had returned to shape my idea of what Sissinghurst might be – not a slightly heartless, controlled and driven visitor attraction but a place in which the people who lived and worked there felt at home, which wasn't rigidly and drearily proper in the way it went about its business, but left some room for the quirks and wrinkles which an unpolished life might give rise to. If old Sissinghurst gave Perch Hill a vision of integrated completeness, Perch Hill could give Sissinghurst a dose of vitality and delight.

It took about five years for this to happen, five years which began in upset and resentment but which by the summer of 2009 seemed, in the usual strangely sudden way of these things, to have settled into a new path. A new farmer, John Hickman, had arrived. He had been by pure chance one of Simon Bishop's pupils, and he brought his Sussex cattle and Romney sheep to the Sissinghurst farm. Nearly a mile of new hedge on old lines had gone in, a big new orchard, new hay meadows, a new vegetable garden, new woods, new wetlands for the willow warblers and nightingales. Above all a new sense of purpose had emerged there, a vision of the place which went beyond curating the lives of Vita and Harold and started to see the 30-odd years they had lived there as just another episode in the long and evolving story whose roots had begun at the end of the Ice Age 10,000 years ago.

That beautifully long perspective, combined with the surge of energy from new people coming to work there and find new purpose there, was a stimulus for Sissinghurst, whose atmosphere rose into a balloon of optimism. For me and Sarah, it was a kind of cure. I felt Sissinghurst lifting away from me. It was set on a good track, with some good people steering it, and Sarah and I felt released by that to turn our attentions again to Perch Hill.

But there is no arrival. As I was writing these pages, we were hit with the news that Simon Bishop, driving home one evening in November 2009, was killed when a deer came in through the windscreen of his car. It was a freak of an accident. Simon never drove fast, nor took any risks. It was in his nature to make life steady and calm. It always seemed in his presence that not only would life be all right: it would be fun and funny. His attachment to the land and its animals, and the constancy which they imposed, meant that thrills and spills were nearly absent from his life. This was blind chance: a car coming the other way down the B road between Plumpton and Netherfield hit the deer and flipped it up into the air and into Simon's car just at the second he was passing.

His death cut holes in all of us. Six hundred people came to his funeral in the big, stalwart Norman church at Battle. Hundreds of pupils and colleagues crammed in there. There wasn't room for everyone to sit down. Colin Pilbeam cut boughs from the Perch Hill wood. With Jo Clark, Simon's sister-in-law,

Sarah made arrangements in the church, mixing the autumn leaves of the oak and silver birch with giant white lilies and laying a wreath of lilies on his coffin.

He had been the centre of so many lives that Perch Hill without him seemed at first like a car whose axles had been removed. Everything was still in place, but nothing was whole. For Tessa and their children, the bleakness was all enveloping. For the rest of us, we had to understand how to go on without him. We had to ensure that the farm network he had set up and run, and the courses he had established at Ivyland and whose pupils had done so much at Perch, would both continue. All of us felt we had a duty to maintain his vision of the furture. He *knew* that Sussex was wonderful. It could only be up to us, and the hundreds of others he had inspired, to make sure it was. At the moment things look fine; his death has brought no break in the systems he worked to establish. There's no arrival; just going on in the best possible way, as Simon would have wanted.

We had always been full of plans for the buildings at the farm, but never had the money to carry them out. There were times when we couldn't buy furniture for the rooms. And because we couldn't afford to do what we wanted to do, we ended up doing nothing. The result was that the buildings, over nearly 20 years, started to look very dog-eared, crumblingly charming as the back-drop to a garden, but scarcely the right way to treat the place you were meant to love. You can be sure enough that 'the messuage called Perchhowse' that was here in the 1580s would not have been crumblingly charming.

Sarah and I saved up enough money to make an investment in the buildings. We didn't want the place swished up. The barn would be made good but remain a barn. The farmhouse itself would be made comfortable, wind- and rat-proof at last, but nothing hi-luxe. We must choose simplicity, the true materials, no gilt downlighters, no knocking down of ancient walls. We shouldn't change the footprint of the house. It should remain in effect as simple as the place required. This was never going to be an estate agent's dream. Instead, we thought, we should put the tiles back on the barn which it had lost in the hurricane of 1987, when the whole roof structure was blown off and put back up there by one of the Weekes boys using the front loader of the tractor.

As the children grow up and begin to earn a living and our expenses begin to drop, we might be able to bring Sarah's school to a gentle conclusion. She and I could then work in there, she in the greenhouse, me in the plain white room where, long ago, I did the lambing in the raw east wind coming off the Channel at night, freezing the water in the buckets around me. We have put up a cow shed, steel-framed and practical. Its floor, as they often used to be, is of rammed chalk, sealed with sour milk. It's a beautiful thing in the winter, an enhancement to the landscape, smelling sweetly of the cattle and their straw. The sheep have their lambs in there once the cattle have gone out to their summer pasture. More than that, though, I have the feeling that this shed somehow pegs Perch Hill Farm down into its place. Even the knowledge of the animals there in the winter, the sight of their hay stacked from midsummer onwards, the late-summer arrival of the straw from an arable farm and then the dunged straw going back on to the fields in the springtime – all this is a deeply rewarding contribution to the tangible cycle of life to which this book and this whole story have been committed. It provides the bass note, the ballast, for everything else here.

I know this bit of country now. The real pleasure is not in the management, control and decision-making that owning land involves. It is something both less and more than that. It is the ability to roam in your mind across the surface of a place which is so well known to you that it has become indistinguishable from who you are. A deeply and properly known stretch of country clamps itself on to your existence like a second skin. In that marriage of you with your surroundings, something extraordinary happens. You can run your mental fingers over the place you know, feel the familiar grooves and hollows, the shiny, well-rubbed parts like the burnished wood on the arm of a chair, or the nicks and elbows in it, the knobbly interruptions.

If you lie in the bath and think of these things, the mind becomes like another eye or hand itself. These are the workings of the sensuous memory. It can feel a remembered landscape like a shepherd at the market feeling along the back of a lamb for the meat covering the bones. Once you know what to look for, and once that sensuous knowledge has registered, it enters a deep and animal layer of the consciousness from which physical shapes can be recalled far more easily than anything that is more refined. It is that instinctive belonging which this book has been about.

The things we have done to the farm and garden are somehow joining us in that state of knowledge. The perennial plants are ageing in the garden and it is acquiring a more serious and adult air, the froth and glitter of the annuals finding a big responsive background in the clematis and roses, which now thicken each summer on the walls. The hedges we have put in are now tall and dense. Many of the trees we have planted have started to acquire a dignity and presence of their own. Young oaks and ashes are pushing up above the hedgerows. Clusters of hornbeams in the Cottage Field can now provide shade for the cattle in the summer. Time horizons are starting to stretch away. What was once a jittery lurch from one month and then one year to the next has now started to acquire the contours of a lifetime's work. I can see Perch Hill now as I will when I am dead, and I think how lucky I am to have lived this life, even to have been through the travails with which this book began, because they were what led me to the rewards Perch Hill held, waiting and in secret.

I know this now: when we spend all our money on a wrecked farm; when we feel drawn to the riches of an ancient landscape; when we hang on to the privacy of a secluded place; when our hearts open to the beauty of a dawn in that place, or a long summer evening; when you are there with someone you love those are the moments when you know the value and richness of being alive.